John W Masury

The carriage painters' companion

also practical hints and directions as to the best mode of applying the same

John W Masury

The carriage painters' companion
also practical hints and directions as to the best mode of applying the same

ISBN/EAN: 9783741119088

Manufactured in Europe, USA, Canada, Australia, Japa

Cover: Foto ©berggeist007 / pixelio.de

Manufactured and distributed by brebook publishing software
(www.brebook.com)

John W Masury

The carriage painters' companion

THE CARRIAGE PAINTERS' COMPANION.

Wm. J. Read, Steam Job Printer, 116 Fulton St., New York.

THE

Carriage Painters'

COMPANION,

WITH

SAMPLE CARDS

OF ALL THE COLORS USED IN CARRIAGE WORK·

ALSO

Practical Hints and Directions as to the Best Mode of Applying the same ;

AND

THE PRORPORTIONS FOR MAKING THE BEST GROUNDWORK FOR LAKES AND CARMINES.

WITH OTHER USEFUL AND GENERAL INFORMATION.

THE RESULT OF

25 Years' Practical Experience in the Paint Shop.

Published by JOHN W. MASURY.

TABLE OF CONTENTS.

CONTENTS.

To Carriage and Car Painters, Greeting.

OFFERING to the Trade this Book of Samples, with hints and directions, we think it not unreasonable to ask of those into whose hands it may come, a proper appreciation of the large outlay of time and money necessary to its production—its entire novelty and general utility. The inventor has succeeded where hundreds have failed! The attempts to produce ground colors of an impalpable fineness, together with the indispensable property of drying quickly and flatting perfectly, have not heretofore met with any measure of success; and this want of success has been due to the absence of machinery perfectly adapted to the desired end. The writer, after many years of experimental trials, has produced a machine whereby the friction—necessary to the pulverizing to absolute fineness of the hard, unyielding substances of which many of the pigments are composed—is possible, without the heating, which has always heretofore been an accompaniment of the process. The remedy was simple and natural, and we may wonder that the application was not made until this late day. It consists merely in applying cold water to the grinding surfaces of the mill in such a way that the running stream shall carry off the heat which must come from extreme friction, and which must otherwise be imparted to the materials subjected to the process of grinding. The difficulty of heating being obviated, there remained only to ascertain by experiment what certain vehicles are best suited to wet and to keep in a useable form the various kinds of pigments employed in the operation of coach and carriage painting; because of the fact that what are entirely suited to some paints are quite as unsuited to others, and the indispensable property of drying quickly could not for a moment be lost sight of. These experiments have been costly, both in time and money, to say nothing of consequent vexations and disappointments; but the entire success which has crowned our labors has in no small degree repaid the expenditure, and it is with entire confidence we offer to the Trade, our ground colors, samples of which—imperfect, of course, from the nature of the material on which they are painted—we herewith present. As each color will be accompanied with a full description of its peculiar properties, with suggestions as to the best mode of applying, both as to ground-work and finish, it is deemed superfluous to attempt a general description of the same, except to assure all consumers that we offer them colors in a shape best suited to the requirements of the Trade. The hints and directions we think may prove lessons to some, and we trust the most skillful craftsman may recognize the truth and importance of what we have said in the pages of this little book

JOHN W. MASURY.

New York, January, 1871.

DSI

7

Coach Painters' Superfine (Jet) Ivory Black.

This color we place first on the list, because it holds the most important place among the various pigments in the carriage paint shop, being more used in finishing coats than all the other paints.

It may be said of this black that beside it all the other blacks are gray. It is ground to such a degree of fineness that a single coat, applied with a soft brush, will perfectly cover a surface of the finest French white china or porcelain without in the slightest degree roughening the same. It will dry ready for varnish in about half an hour, and some of the most skillful coach painters in New York, and elsewhere, have declared that its working under the brush is a perfect charm. Its economy over black produced in the paint shop is not a question.

This Black is put up in our patent cans in quantities of one pound and upward, and will keep soft and fresh for any length of time, even after being opened, if kept covered from the air.

PRICE PER POUND IN ASSORTED CANS, - - - - .50
" " " " 100 ℔ LOTS, LARGE, do. - .45

N. B.—This color will dry ready for varnish in half an hour or less when thinned with clear turpentine. By using a portion of raw oil it may, of course, be made slow to suit the work.

JOHN W. MASURY'S
Superfine Colors.

PREPARED ESPECIALLY FOR COACH AND CARRIAGE PAINTERS,
AND ORNAMENTAL AND CAR WORK.

Coach Painters' Superfine Ivory (Jet) Black.

VALENTINE'S VARNISHES.

THE American Institute Fair, at their thirty-ninth annual display of 1870, gave to VALENTINE'S VARNISHES the award of "*Honorable Mention.*" The value of this award is explained in the judges' report, thus : "The Honorable Mention award is in most cases intended to indicate that *the article so praised marks an important step forward in the discovery of things generally useful.*" In the report of the Committee on Paints and Varnishes, they add : "These varnishes, for body, brilliancy and durability, are *the best*, in our judgment, *ever manufactured in this country.*"

VALENTINE & Co. have labored long and perseveringly to improve the manufacture of coach varnishes, and to compete with the imported article by an American varnish equally good ; and they are now able to offer a sufficient guarantee that this result has been accomplished. The fact is beginning to be generally accepted, and the report of the American Institute is confirmed daily by "honorable mentions " awarded them by *practical carriage painters* located in all parts of the United States. The general expression of their report is this : "*Valentine's Coach Varnishes are fully equal to the best imported in every respect.*"

VALENTINE & CO.,
88 CHAMBERS STREET, NEW YORK.

THE HUB

A Monthly Magazine Devoted to Carriage Building.

Address: EDITOR OF THE HUB,

88 Chambers Street, New York.

In March, 1871, "The Hub " and the "New York Coachmakers' Magazine " will be combined, forming a Magazine of 24 pages. It will be divided into the following departments : 1. Wood Shop ; 2. Smith Shop ; 3. Paint Shop ; 4. Trimming Shop ; 5. Office ; 6. Correspondence ; 7. Trade News. Price, $3.00 per year, beginning with April, 1871.

How to Paint a Carriage.

THERE are many ways of proceeding to the same objective point; and doctors even will disagree as to the proper mode of treating the same symptoms. Coach painters can hardly be supposed to be more unanimous than M.D.'s, particularly when the latter fraternity are leagued by all sorts of oaths and bonds not to affiliate or hold consultation with a School of Medicine, which proposes to kill—or cure—by some irregular method.

No doubt some of my fellow craftsmen will see a better road than I propose to travel to reach the same point; which is, a well painted job in every respect.

The writer does not belong with that class which takes for granted, that a thing is good *because* it is new; nor with those who cling to a time honored custom, simply for the reason that the same is sanctioned by long use; nor, with those who believe that any particular theory or mode of procedure includes all that is good and avoids all which is bad.

The prejudices of craftsmen are difficult to meet and hard to overcome. One clings to a system and dogmatically pronounces it best, simply because he has never tried any other and he hates innovation. Another readily adopts, or tries, all suggested improvements, and becomes an innovator because of the charms and excitements of novelty. As a rule the first will win in the race: but the second is useful in his day and generation.

Without attempting to trace the progress of improvement

[*Continued on page* 15.]

Coach Painters' Superfine Ivory Black.

This color is made from the best English Drop Black, and is ground equally fine with the "Jet" described in the foregoing page. All the remarks applied there to the "Jet" Black are equally applicable to this, as to drying, ease of working, etc. It is put up in the same way, and is sold —

ASSORTED CANS, - - - - - - - 45 cts. per lb.

100 ℔ LOTS, LARGE CANS, - - - 40 " " "

The sample shown on the opposite page was painted by the following process: About four o'clock, P. M., a coat of Black, thinned with turpentine and a very small quantity of raw oil, was applied to wrapping paper, stretched on a wooden frame. At 10½ on the following morning, the rough surface of the sheet was rubbed with fine sand paper, and another coat of black, thinned with clear turpentine, was applied. In two hours this was ready for varnish.

Coach Black.

This Black will be found superior, both in fineness and color, to any of the blacks in common use. For drying see remarks above. Price in

ASSORTED CANS, - - - - - - - 40 cts. per lb.

100 ℔ LOTS, LARGE CANS, - - - 35 " " "

Coach Painters' Drop Black.

This Black has been sold by us for the past twenty years, and is too well-known among consumers to require a particular description.

JOHN W. MASURY'S
Superfine Colors.

PREPARED ESPECIALLY FOR COACH AND CARRIAGE PAINTERS,
AND ORNAMENTAL AND CAR WORK.

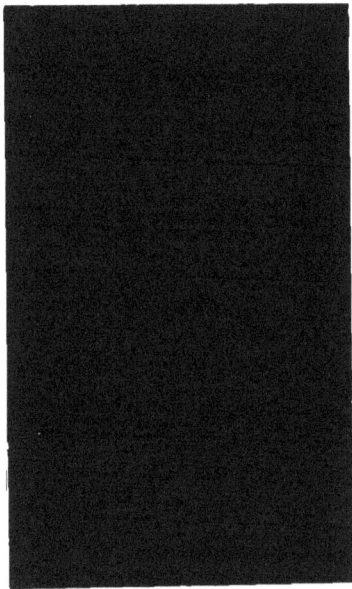

Coach Painters' Superfine Ivory Black.

MURPHY & COMPANY,

Varnish Makers,

53 & 55 OLIVER STREET,

NEWARK, N. J.

PRICE LIST

Of their Goods adapted to the uses of CARRIAGE BUILDERS and RAILWAY COMPANIES.

WEARING BODY, For Outside Finishing. - - - - $5 75

This Varnish is very pale, of the freest working properties, and has unequalled durability. It is especially adapted for last coats on exteriors of Railway Cars and Coach Bodies, where the greatest durability is required. In the cooler months it can be employed with safety in all ordinary shops.

MEDIUM DRYING BODY, For Outside Finishing. - - - $5 75

This Varnish is fully equal to our Wearing Body in paleness and working properties, and is intended for use on all work requiring moderate dispatch, when time cannot be given for our Wearing Body to harden, especially during the summer months. This Varnish is also employed with great satisfaction for last coats on Engines.

HARD DRYING BODY, For Under Coats, &c. - - - - $5 00

This Varnish is also equal to our Wearing Body in paleness and freedom of working, and is mainly intended for under coats, for preparing a surface of great durability for our finishing Varnishes. It is also used with great satisfaction for last coats on interiors of Cars, giving a brilliant and durable finish.

RUBBING BODY, For Under Coats. - - - - - $4 50

A pale Varnish sufficiently free in its working properties to permit its employment on the largest panels. It is intended for under coats, where moderate dispatch is desirable.

ELASTIC CARRIAGE, For Last Coats on Running Parts. - $5 00

An Elastic Varnish of great durability, intended for finishing coats on running parts of fine carriages.

HARD DRYING CARRIAGE, For Last Coats on Running Parts. $4 50

This Varnish has heavy body and is intended for use on running parts where it is desired to finish with one coat on flat color. It is also a very desirable Varnish for use on old work.

COACH MAKERS' JAPAN, For Binding and Hardening Paints. $2 00

A Superior dryer for Coach and Car Painters' uses. It is made from finest Shellac and will mix readily with oil.

14

in vehicular construction, from the rude log-wheel carts of the ancients to the graceful and elegant vehicles of the present day, it may be asserted without fear of contradiction, that there are few things in our advanced civilization and refinement which are more attractive, which combine more fully the useful and the beautiful, than the gracefully modeled, luxurious and comfortable carriages which are turned out from the first-class city and country manufactories.

To paint a carriage in the highest style of the art requires a judgment matured, an eye to appreciate combinations and contrasts, and a hand cunning and skillful to execute and perform. In nothing more than this, is it true, that practice alone makes perfect. Written rules and directions are valuable only as hints and suggestions, which, if properly heeded and carried into practice, may lead to the correction of errors, which exist because of the want of proper instructions. As well might one expect to educate the ear to harmonious combinations of sounds by a treatise on musical composition, as to teach the art of painting by mere words. Yet, while the finished workman needs no written rules, there are many throughout our country, living remote from the great centers of population, who profess and practice the art of carriage painting, without the opportunity of perfecting themselves in the higher branches of the profession. In the hope that to such, our directions may prove of practical benefit, we proceed to give the mode of proceeding in the old method of carriage painting. Of the new and shorter method we shall treat hereafter.

As the priming, or first coating of the new wood, initiates

[Continued on page 19.

Raw Umber.

This pigment is a native ochre, and occurs in the island of Cyprus. It is known in the Trade as Turkey Umber, although a great portion of the Umber sold in this market is a native product. Genuine Turkey Umber is a soft, brown pigment, transparent in oil, and abounding in manganese, from the presence of which it derives its drying property. It is one of the most useful colors in the stock of the house painter, and is much used in graining, and in producing, with white, pure quaker drabs and browns. With blue, it affords a good neutral green, very permanent.

The American substitute for this pigment seems to possess none of the properties of the genuine article, except a resemblance in color.

In assorted cans, 1 ℔ and upward, 35 cts.

Raw Umber, with White and a little Chrome Yellow, makes a soft, delicate shade of yellowish drab.

N. B. We grind the genuine article only, and purchasers may depend on getting the best of its kind, under all circumstances.

John W. Masury's
Superfine Colors.

PREPARED ESPECIALLY FOR COACH AND CARRIAGE PAINTERS.
AND ORNAMENTAL AND CAR WORK.

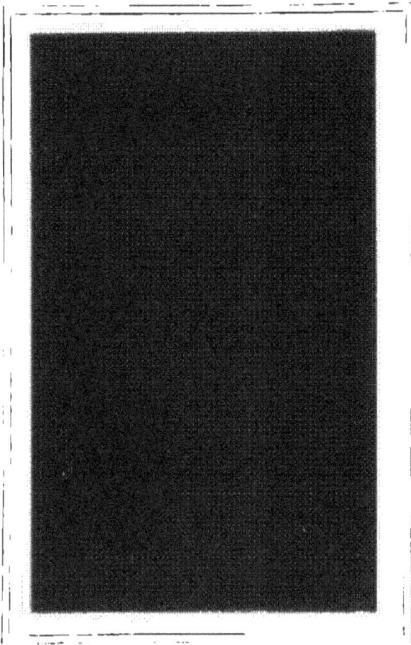

Superfine Raw Umber.

ESTABLISHED 1827.

EDWARD SMITH & CO.,

LATE

SMITH & STRATTON,

Manufacturers of

VARNISHES,

No. 161 WILLIAM STREET,

Edward Smith,
John A. Elmendorf.

NEW YORK.

Wearing Body,

Rubbing do.

Color & Varnish,

Brown Japan.

PREMIUM VARNISH.

1827.

E. S.

TRADE MARK.

Car Body,

Do. Rubbing,

Do. Flowing,

Gold Size.

18

the operation, that simple process requires a word or two at the start. First, as to what shall be the material used ; and second, how to apply it ; and these are important questions, as the durability of the job depends in no small degree on the soundness of the initiatory proceedings. It will not be denied that whatever material adheres most tenaciously to the wood, best resists the changes of temperature, dryness and dampness, and wear and tear is the best, whether it be white lead and raw oil, or boiled oil or Japan on wood-filling, or any other substance. (Of the use of permanent wood-filling we design to speak more at length by and by.)

To mix the priming coat, thin a small quantity of ground white lead with raw linseed oil, adding a few (say two or three) spoonsful of Japan or Japan gold-size, for a dryer, and enough turpentine to make the paint work easily. Apply an even coat of this paint with an ordinary bristle paint brush, taking care to work the color well into the nail heads, crevices and corners of the body, wheels, and carriage part. After the body has stood for four days for drying, the carriage part being meanwhile in the blacksmith's shop, undergoing the process of ironing, mix color for second coat as follows : dry white lead mixed stiff in Japan and raw oil, equal parts, and ground through the mill. Thin to proper consistency, with turpentine, and apply with an evenly worn brush, taking care to work the color down smoothly. This coat should stand four days for drying, and hardening; after this, fill all the holes, crevices, chinks and imperfections in the wood, with hard putty, made thus : white lead three parts, whiting one part, wet with a mixture of two

[Continued on page 23.

Burnt Umber.

To produce this most important pigment, the crude Umber is put in iron retorts, and subjected to a heat more or less intense. The result is the changing of the tone of color to a very much deeper and more red brown. The drying property is also increased by burning. No color in the stock of the painter is better known and appreciated than this. It is much used for graining and for producing, with white, warm clear browns and stone colors. It is transparent and permanent. The Umber pigments require to be very carefully prepared and finely ground, as it is only by reducing them to the last degree of fineness that they show their real colors.

In assorted cans, 1 ℔ and upward, ℔ ℔, 35 cts.

This color will dry in about two hours.

Burnt Umber, with White and Orange Chrome Yellow, will give a variety of shades of clear, warm drabs.

Burnt Umber, with White and Lemon Chrome Yellow and Scarlet Lake, will give a rich shade of tan color.

JOHN W. MASURY'S
Superfine Colors.

PREPARED ESPECIALLY FOR COACH AND CARRIAGE PAINTERS.
AND ORNAMENTAL AND CAR WORK.

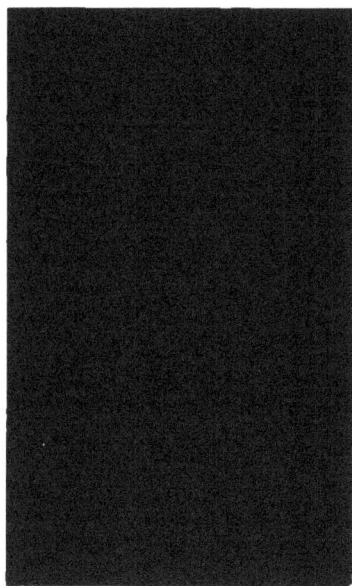

Superfine Burnt Umber.

parts linseed oil, two parts varnish, and one part Japan or gold-size. When filling the screw-heads and other hollows, allow the putty to stand a little above the surrounding parts ; that is, the holes should be more than full to allow for any possible shrinking. All open-grained wood, as ash, must be filled with soft putty, made of white lead wet with equal parts varnish and Japan, using a square-pointed putty knife; care must be taken to fill all the pores of the wood, and thoroughly remove all superfluous material from the surface. Let the body stand three days; at the end of which, apply the second lead coat, mixed dry lead in three parts Japan and one part oil; mix stiff, reduce with turpentine, and apply as before. Observe that care should be taken to spread every coat evenly, whether it be lead, roughstuff or color. This should stand three days before the application of the third and last lead coat, which should be mixed dry lead, wet with four parts Japan and one part oil. After two days (four is better if not pressed for time), the body is ready for roughstuff. We can suggest no better mode of mixing roughstuff than the following, viz.: Two parts English filling, two parts dry white lead, wet with mixture of two parts varnish and one part each Japan, oil and gold-size. Make into stiff paste and reduce with turpentine to proper consistency for spreading with a well-worn brush This should be allowed two days for hardening before the application of the second coat, which should be mixed in one-half the quantity of oil used in the first coat. The following day the third coat, in which no oil should be used, may be applied, and again the next day, the fourth coat, which should be mixed the same as the third coat ;

[Continued on page 27,

Sienna.

Terra-de-Sienna is a ferruginous native pigment of yellow brown hue, producing, with white, bright sunny tints. That known in commerce as Italian Sienna, is the most esteemed, and is, in fact, the only article bearing the name of Sienna which possesses any real value as a pigment for fine painting. The so-called American Siennas are vastly inferior, both in color and transparency. Crude or Raw Sienna, when subjected to a high degree of heat, loses its yellow complexion and takes on a deep, clear brown-red hue, retaining, at the same time, all the transparency of the unburnt material. It is an invaluable pigment, and is extensively used in every department of painting. By admixture with Roman, or Yellow Ochre, or Raw Sienna, or any other transparent yellow, and Antwerp or Indigo Blue, it affords fine Olive Greens.

Burnt Sienna is a rich, transparent brown orange. It has strong coloring properties, and is permanent to the last degree.

Price, ℔ lb. in 1 and 2 ℔. cans,　-　-　-　　37 cts.

N. B.—Raw and Burnt Sienna are both the same price.

JOHN W. MASURY'S
Superfine Colors.

PREPARED ESPECIALLY FOR COACH AND CARRIAGE PAINTERS.
AND ORNAMENTAL AND CAR WORK.

Superfine Burnt Sienna.

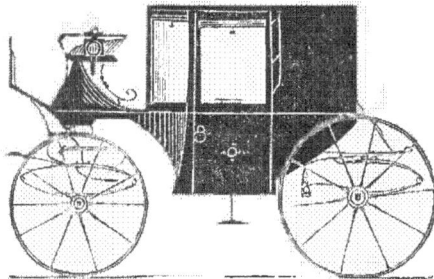

that is, without oil. The roughstuff should, of course, be ground finely through the mill, as should all the other mixtures, into which dry lead enters as one of the component parts. The last coat of roughstuff should be followed by the guide coat, of French yellow ochre, mixed in Japan and turpentine.

The body may now go to the smiths to be hung up. That done, the wood-worker should smooth up all places where the beds may project over the axles, put on bands, etc. The painting process should now be resumed by priming the iron work, which should stand two or three days to dry. While the carriage is hardening the scouring of the body may be proceeded with. This should be done by an experienced hand, as great care is required to prevent the pumice-stone from cutting through the successive coats of paint to the wood. The lump of pumice stone should be kept well filed, and plenty of water should be used to prevent the pores of the stone from becoming clogged with the paint. This process should be continued until none of the guide coat is left; and being completed, the body should be washed off with clean, cold water, using the water tool for corners and all places where the particles removed from the surface by the action of the pumice stone is apt to collect. The body may now be left to dry for twenty-four hours, and work resumed on the carriage parts. First, cut down thoroughly every part with No. 2 sand paper, dust off and apply lead coat, mixed as follows, and ground finely through the mill: dry white lead, in equal parts of Japan and raw oil, reduced with turpentine. Judgment is required in the application of this coat, because if the paint be too thin, the pores of the

[Continued on page 31.

Van Dyke Brown.

This useful pigment is a bituminous earth of vegetable origin. The most valuable kinds are found in Germany. The color is very rich, deep, transparent brown. It is a favorite color with many artists, and is used by the house painter mostly for graining. It is of no value unless very finely ground. Powerful dryers should be used for grinding it, as it is of all known materials the worst color to dry. It is clear in its pale tints, and deep and glowing in shadows. Very permanent.

With white. it gives a clear, warm gray, of a tint which no other single color, with white, will produce.

Price, ℔ lb, in assorted cans, 1 ℔ and upward, 45 cts.

Silver White, or White Lead, with Yellow Lake, or Dutch Pink, with a little Lemon Chrome Yellow, will give a very rich shade of olive drab or amber color.

JOHN W. MASURY'S
Superfine Colors.

PREPARED ESPECIALLY FOR COACH AND CARRIAGE PAINTERS.
AND ORNAMENTAL AND CAR WORK.

Superfine Van Dyke Brown.

wood will remain unfilled; and if too thick it cannot be spread evenly. Apply with bristle paint brush, working the paint well into the wood.

This coat should stand at least four days for hardening; but, in the mean time, it would be well to soft-putty the rims, faces of spokes, and all the flat surfaces of the carriage part. Putty for this work ṣhouḷḍ ḅe ṃaḍe of dry white lead, wet with equal parts ọ̣f ..., using a square-bladed putty knife ... the grain of the wood, taking care ... on the surface, because any loose pẹ ... mble and fall away after the carriag ...

Returṇ ... work is resumed on that by going liḡ ... .ce, with the very finest sandpaper uṣ ... alar pains must be taken to clean oṛ ... uld any imperfection be discoverẹ ... n unfilled, the same must be stoṛ ... ẓe body will be ready for color.

It is proḷ ... .ıse with the old, and it seems to us, unnecessary custom, of going over the work again with what is called the surface lead coat.

It will now be understood that the successive coats of paint, with the labor of rubbing and smoothing, have brought the surface to the best possible condition for receiving the first coat of color. This surface, which has been gained by the expenditure of so much time and labor, it should be the constant effort of the workman to preserve, because, for a scratch or in-

[Continued on page 35.

Royal Red.

This not brilliant, but useful and desirable pigment, is an artificial ochre, composed mostly of iron ore. It produces very good Browns in combination with Blue or Black, and used by itself gives good groundwork for either pure Vermilion or its counterfeit, the so-called American Vermilion. It is entirely permanent in col

Price

The varnish   evening
previous to finis.

Do not use   varnish
room, as the vap   ng   .eleterious
effect upon the v.........

Proprietors should never take strangers into the varnish room during the operation of finishing.

An even temperature should be kept in every varnish room. About 68 ° or 70 ° is most favorable, all things considered.

Body varnish-brushes should not be used on carriage parts.

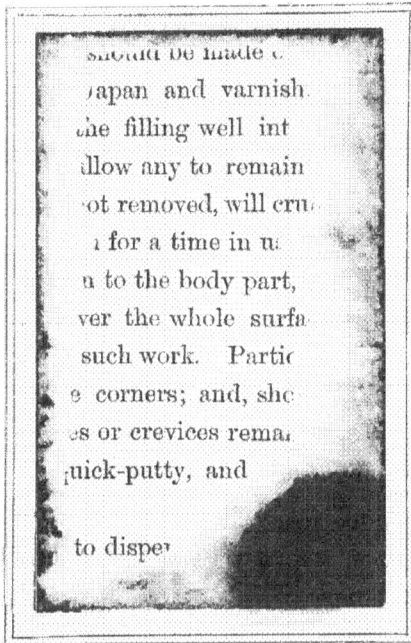

dentation on the same, there is no remedy but to go half-way back and begin again.

It is proposed to finish this job in black; that being the most common, as well as the most important of all the colors used in the carriage paint shop. It does not take long to learn that black (which is the carbon resulting from the burning of animal bones in close vessels), is serviceable and valuable, just in proportion to the minuteness of the division of the particles. Black, not finely ground, has little body, and comparatively little adhesive property. The ordinary appliances and means for grinding colors in the paint shop are not equal to the task of grinding black to that degree of fineness which is essential to produce the best effects in finished black work.

Nor has there been, either in this country or abroad, until our recent invention, any machinery whereby hard pigments, like Black and some of the Lakes, could be reduced to that impalpable fineness, on which their value and good working qualities mainly depend, without adding so much to the cost as to put them beyond the use of coach painters entirely. Asking pardon for this digression, and taking for granted that you have on hand a stock of MASURY'S Superfine Colors for Coach Painters' Use, and that the body, which was left ready for color, is to be finished in the best style, the next proceeding is to open a one pound can of Ivory "Jet" Black, which will be done in a second with the help of a penknife blade. This black will be found finer than, one year ago, it was thought possible to reduce any substance, and so soft and manageable that it incorporates at once with the thin-

[Continued on page 39.

Indian Red, Superfine.

This color is of all the most common and the most held in esteem in the carriage paint-shop. It is a pure per-oxide of iron, of a rich laky, red brown color, entirely permanent and of astonishing body and covering property. It is hard, gritty, and yields very reluctantly to the process of grinding. Indeed, to reduce it to a perfect fineness, equal to that shown in the sample, would be quite impossible in any ordinary paint-mill, or with slab and muller. The cost of painting with our Superfine Red is next to nothing, as a quarter of pound will cover a surface of about eight square yards, making the cost per yard about one cent. To produce a pound of such color would (supposing it were possible in the paint shop), be very much more than we charge for it.

Superfine Indian Red, ℔, - - 40 cents.

The smoke from bituminous coal will cause varnish to look blue and cloudy.

JOHN W. MASURY'S
Superfine Colors.

PREPARED ESPECIALLY FOR COACH AND CARRIAGE PAINTERS,
AND ORNAMENTAL AND CAR WORK.

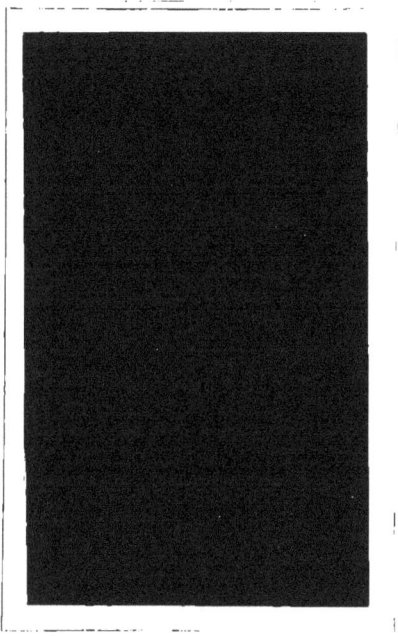

Superfine Indian Red.

ning, and the mixture becomes as homogeneous as though it were all one substance. Enough of this black to go over the work is taken from the can and thinned with turpentine, using a little oil and varnish if you have time. This black will be found to cover whatever it touches, and a solid job with one coat may be made over pure white. Put on with flat camel hair brush, which leaves no brush mark. This coat had best stand one day before the second coat of black is applied. That done, the work is ready for first coat of varnish. This is an important point in the process of our work. So far, all has been done with reference to durability as well as beauty, and as a coat of bad varnish will nullify all that has been done in that way, it behooves us to be not a little particular about the matter. It is not for us to say who makes the best rubbing varnish ; but we have no hesitation in saying, what, in our opinion, a rubbing varnish *should be*, to fulfill all the requirements of the occasion. It must flow smoothly : it must dry hard, and yet elastic : it must rub well, clean down well, and not sweat. If you can find a varnish fulfiling all these necessary conditions, no matter what name it may bear, apply a coat of it to the work in hand : not a heavy coat, but a light one, with a flat brush, of which there are several kinds intended specially for varnish. (For a most complete assortment of brushes for coach work, we call attention to the establishment of HENRY W. GEAR & CO., New York. The manufacture of Fitch, Badger, bristle and other serviceable brushes for carriage work, is a specialty with them ; and, in the way of tools, the painter cannot be wrong if supplied with a stock of brushes of their make). A thick, flat Bad-

[*Continued on page* 43.

Tuscan Red.

This fine rich Lakey color is of the same nature as Indian Red. It has a dense body, dries and flats well, and is entirely permanent. The sample shown on the opposite page is one coat of Tuscan on a ground of Indian Red and one coat of clear varnish. As a ground for Munich, Carmine, and Purple Lakes, and for deep Carmine, it has no superior.

Tuscan Red, in assorted cans, ℔ 75 cts.

To make a rich shade of Bismark Brown take a ground of two parts Burnt Umber to one of White Lead. Put over this two coats Burnt Sienna and glaze with Bismark Brown, made as follows: one ounce best gold bronze, half ounce of Carmine and half ounce of English Crimson Lake. When a very light color is required use English Vermilion for ground in place of that named above.

N. B.—That most of so called Siennas and Umbers sold in the market are only imitations of the genuine. To make good work requires the best colors. Our superfine colors are selected for us abroad and are warranted in all cases to be the very best.

JOHN W. MASURY'S
Superfine Colors.

PREPARED ESPECIALLY FOR COACH AND CARRIAGE PAINTERS,
AND ORNAMENTAL AND CAR WORK.

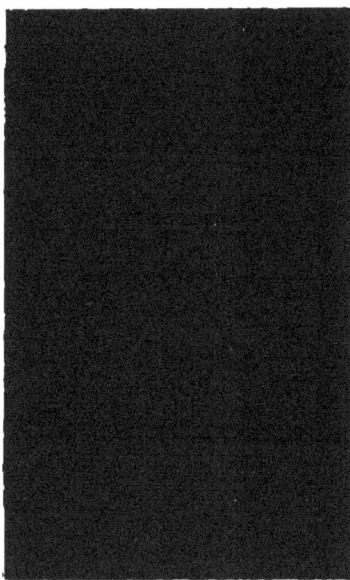

Superfine Tuscan Red.

M. H. SCRIBNER,

MANUFACTURER OF

TRANSPARENT

DRYER,

For Coach, Car and House Painters' Use.

Office : 23 BANK STREET,

NEWARK, N. J.

THE subscriber would respectfully call attention to his TRANSPARENT DRYER, manufactured with great care, with special reference to the wants of Car and Coach Painters.

An experience of several years has given him a thorough knowledge of this branch of the varnish business, and he trusts that all those who are in need of a good, substantial dryer, may have sufficient confidence in it to at least give his stock, which is made from none but the best material, a thorough trial.

The TRANSPARENT DRYER is a superior Oil Dryer, and its strength is nearly double that of ordinary Japan.

It will not discolor pure white or any fancy colors ; does not thicken up or destroy the fluidity of the paints, which makes it well adapted to use in mixing paint for striping.

It is warranted not to crack, and to incorporate freely with either raw or boiled oil ; is a complete binder of all colors, and is just what is needed by all first-class Car, Carriage. House, Sign and Ornamental Painters.

ger hair varnish brush, of chisel form, about three or three and a-half inches wide, is recommended for such work as is now the subject of treatment. Such a brush, if well cared for, will last a lifetime, and grow better with age. But let us return to the body, which was left with one coat of rubbing varnish, and which must now be put aside to dry for three days. During this time work may be resumed on the carriage : first, by going over it again with sandpaper ; and now care must be exercised not to rub the sharp angles through to the wood. After this, dust off and apply second lead coat, mixed as follows : dry white-lead wet with a mixture of Japan and oil, in the proportion of three parts of the former to one of the latter, and made stiff ; reduce with turpentine and apply as before, observing same directions as to grinding, reducing, etc. After three days, another slight sandpapering, and the last lead coat may be applied. In this last coat no oil need be used, but clear Japan, and the paint should be applied as before. This being the last lead coat, we of course depend upon it for the smooth, perfect surface required for the reception of the color, which, with striping and varnishing is to complete the job. For cutting down this coat use number one sandpaper, and be very careful to smooth out every corner and bead, and around every bolt-head, nut, etc., and remember that the bases of the spokes require attention equally with the centres, as also do the hubs and rims. This operation, simple as it may seem, is no " child's play," and must not be entrusted to a careless hand, as the same amount of rubbing applied to the sharp corners as to the flat and rounded surfaces, will remove all the successive coats down to

[*Continued on page* 47.

Ultramarine Blue.

Until recently the only pure blue available for painting was True Ultramarine, or "Lapis Lazuli," a precious stone, found principally in Persia and Silesia. All attempts to extort from nature the secret of producing this wonderful azure were for a long time unsuccessful, until, at last, a French chemist, Guimet, by analyzing the true substance, was able to produce by chemical agency an imitation so perfect that best judges were at fault in distinguishing the *real* from the factitious article. In parting with her secret, however, Nature seemed to exact almost as much in the way of compensation as the secret was worth. As an oil paint, this beautiful pigment is, of all, the most intractable and stubborn. It is very hard to grind to impalpable fineness, is of all earthy paints the most transparent and difficult to manipulate. Yet, so beautiful is this "heaven's own blue," that, in contrast with it, all blue paints seem to change to green and greenish browns. The sample on opposite page is not pure, but tinted with white. The first coat was our pure Ultramarine and our New White, in equal parts; this made a body color. The second was one-quarter White to three of Blue; and the third coat was seven parts of Blue to one of White. Pure blue might have followed the third coat, and have made a solid job. The Blue we grind is, however, a much more costly material than the Ultramarine usually sold in the shops. It costs us twice more, in gold, than the ordinary Blue is sold for in currency.

Ultramarine Blue in 1 ℔ and ½ ℔ cans, ℔, $1.

This Blue dries in about half an hour.

John W. Masury's

Superfine Colors.

PREPARED ESPECIALLY FOR COACH AND CARRIAGE PAINTERS.
AND ORNAMENTAL AND CAR WORK.

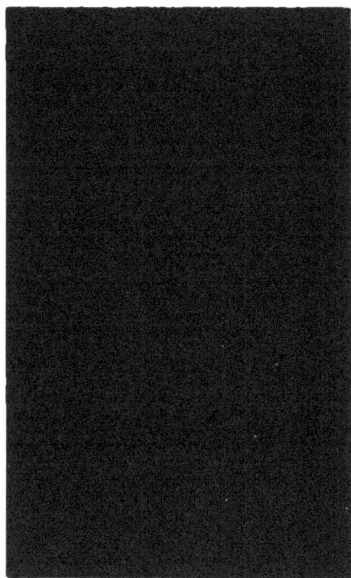

Superfine Ultramarine Blue.

New Haven
WHEEL COMPANY,

218 to 230 York and 2 to 10 Ashmun Sts.

NEW HAVEN, CONN.

Manufacturers of

WHEELS and WHEEL MATERIAL,

OF ALL DESCRIPTIONS. ALSO THE CELEBRATED

Sarven Patent Wheels,

WHICH, FOR

Strength, Durability, Lightness and Beauty,

FAR EXCEL ALL OTHER WHEELS IN THE MARKET.

It is not a new patent, depending upon the success of future experiments to demonstrate its practical value, but has been growing steadily into public favor for the past twelve years   This, with the numerous testimonials we have received regarding its merits, fully suffice to verify all that we or the patentee have ever claimed for it.   It is equally well adapted for the lightest Trotting Wagons or the heaviest Steam Fire Engines.

FOR FULL PARTICULARS SEND FOR CIRCULAR.

SPOKES, Finished and Unfinished, HUBS, RIMS, WHIFFLETREES, HANDLES, &c., &c.

ON HAND AND MADE TO ORDER FROM THE

BEST OF EASTERN HICKORY.

Office: 224 YORK STREET.

HENRY G. LEWIS, President.   EDWARD E. BRADLEY, Secretary.

46

the wood; and, as these parts receive most of the wear and
tear of actual use, it follows that these, of all, require to
be best protected with the paint. The smoothing being properly performed, and *the loose particles removed from every part,
nook and corner*, the work is ready for first coat of color. That
portion of the ground black remaining in the can, after the
painting of the body, will be found—supposing it to have been
kept well covered with turpentine or water—as soft and pliable
as when first opened. Mix a proper quantity of this with turpentine, using oil and varnish at discretion, and apply with flat
camel-hair brush. Ten hours will be sufficient to dry this coat,
when the second will follow, mixed the same as the first coat,
with a little more varnish and less oil, if any. If the work is to be
finished with a very wide stripe, put this on before the first coat
of varnish. The carriage parts being ready for first coat of varnish, apply rubbing varnish, which should be as good in every
respect as that used on the body, and as carefully put on.
Leaving this to harden, return to the body, which was left with
one coat of varnish, and it will be found hard enough for first
rubbing. Provided with a piece of cloth or felt and finely pul-
verized pumice-stone, a water tool, and plenty of clean cold
water, proceed to cut down the varnish as closely as possible, being careful not to go through to the color, and not
to allow the pumice-stone to dry on the varnish; use the
water-tool freely in all the corners and around the mouldings. This operation will be repeated through three successive coats of varnish, and the body is ready for the trimming shop. The carriage part must now be subjected to

[*Continued on page* 51.

English Vermilion, Pale and Deep.

This indispensable pigment, although one of the most important, is really one of the simplest in its constituent parts of all the paints in use. It is composed of two elementary substances, both well known, one of which is of universal distribution. Sulphur combined with quicksilver, in the proportion of about one part of the former to five parts of the latter, under a certain mode of treatment produces the pigment mostly known in the trade as English Vermilion. The test of purity in this paint is simple and inexpensive. A small quantity put on a plate of iron heated almost to redness, will burn for a time with pure blue flame, exhaling a sulphurous odor. After the burning of the sulphur, the mercury will entirely be evaporated by the action of the heat, leaving no residuum, supposing the article to be pure. In painting the sample sheets, specimen of which is shown on opposite page, one pound of ground Vermilion was consumed in giving one coat to about twenty-two square yards of surface. Therefore a coating of pure Vermilion may be had at the rate of about seven cents per square yard. So, a wagon presenting a superficial area of, say, seven square yards may be painted over with this brilliant color at a cost for material of about fifty cents.

Price per pound in assorted cans, - - - $

JOHN W. MASURY'S
Superfine Colors.

PREPARED ESPECIALLY FOR COACH AND CARRIAGE PAINTERS,
AND ORNAMENTAL AND CAR WORK.

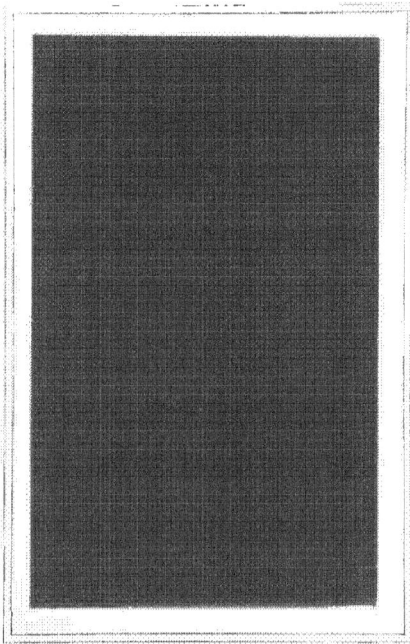

Pure English Vermilion, Light.

50

the same rubbing process as has been applied to the body. This work must not be trusted to unskillful hands. An expert only can do it to perfection. If performed by inexperienced hands, the result will be an untimely striping of all the sharp angles, and the prospect of a well-finished job materially impaired. Supposing this delicate operation to be successfully performed, the striping is next in order. On this subject there is not much to be said. If any one supposes he can do this because he has been told "how to do it," a single attempt will be all sufficient to cure him of his vain delusion ! There is no royal road to this accomplishment ; its attainment is through the steep path of long-continued practice. The striping done and dry, a thorough washing must follow; and be sure that every particle of dust you leave upon the work, will be found by the varnish brush, and carelessness in this respect has too often called down maledictions upon the head of the innocent varnish maker. The carriage parts removed to the varnish room, are ready for finishing coat ; and the writer confesses himself at loss how to give *any hints even* which shall prove of value as to the successful performance of this, of all, the most important in the whole proceeding. A knowledge not only of the nature of varnish generally, but of the particular varnish to be used in the operation, is indispensable to success. To become an adept in this art requires long experience, confidence, and self-possession ; and, we may add, a good conscience. A mistake in this, is little less than a crime ! and your shortcoming will not only rise in judgment against you, but will be known and read of all men.

[*Continued on page* 55.

Vermilion.

True Vermilion is a rich bright color, of dense body and permanent when exposed to the light. This is one of the few pigments which will not bear too much grinding. If crushed to the degree of fineness which some paints absolutely require to develope the color which is in them, it becomes dull, dead and comparatively worthless.

The so-called American Vermilion is made from white lead and chromic acid (bi-chromate of potash). It must not be rubbed so hard as to break the crystals, or it loses its vermilion hue and shows what it really is, a deep orange chrome.

Red Lead is a pure oxide of lead and is produced by exposing metallic lead to the action of the heat, taking care not to fuse it. It is mostly employed as a pigment in painting iron vessels and iron work generally. It has an affinity for the latter metal and serves admirably in protecting it from corrosion.

Orange Mineral is another name for Red Led and is produced by the slow calcination of White Lead in iron trays.

JOHN W. MASURY'S
Superfine Colors.

PREPARED ESPECIALLY FOR COACH AND CARRIAGE PAINTERS,
AND ORNAMENTAL AND CAR WORK.

Pure English Vermilion, Deep.

The body received from the trimming shop, is ready for rubbing, preparatory to the finishing coat of varnish. This, too, is a delicate piece of work, and requires judgment, skill and patience. Remember that a *mote* on a panel becomes a *beam* in the eye of the beholder, and the smallest speck looms up like a distant hill in a misty atmosphere. Having completed it (for better or worse), close the door reverently behind you, lock it, call on your good angel to protect your work from harm, and await the result.

If not pressed for time, it will be well to allow the body to stand over one night before finishing. Remove it to the finishing-room, which was put in order the previous evening, wash it off thoroughly with cold, clean water, using a clean sponge, and a chamois skin which has been well broken in. Do not use dusters which have been used on lead or color, or the mouldings will be discolored. After dusting off well, take a dry, flat Fitch brush and wet the ends of the hair with a small quantity of varnish. Let this stand for half-an-hour, and then go carefully and lightly over the whole surface. This will pick up every remaining particle of lint and dust and there remains only to apply the varnish. This should be done as you should say your prayers, without the presence of any third party; and being done, retire without ostentation, locking the door behind you, and keeping it locked, until the surface is no longer liable to injury from dust.

The next thing in order is to care for the tools. The brush used for picking up the lint should be first softened with a little oil and then thoroughly washed with soap and water, and

[*Continued on page* 59.

Green Pigments.

Greens, so abundant in the vegetable kingdom, are rare in the mineral world, copper being the only metal which gives in its combinations the various shades of green in common use. All the greens used in painting are either copper greens or chrome greens.

Chrome Green (when moderately pure) possesses a dense body, or covering property, and belongs among what are known as body colors. These colors are not as permanent as the copper greens, being compounded of blue and yellow, which colors are not affected in like degree by the action of light. The copper greens, however, are transparent, but far more durable and retain their brilliancy much longer when exposed to the action of the sun's rays.

Paris Green, or, more properly, Scheele's Green (arsenite of copper), is a comparatively late discovery and was first manufactured by the chemist whose name it bears. It consists of about twenty-eight parts oxide of copper and seventy-two parts arsenic. It is one of the highly transparent pigments, will not bear grinding, works badly under the brush and is by no means a favorite with the painter.

Coach Painters' Green, Light. Price per pound, in 1℔ cans and upwards, 50 cents.

JOHN W. MASURY'S
Superfine Colors.

PREPARED ESPECIALLY FOR COACH AND CARRIAGE PAINTERS,
AND ORNAMENTAL AND CAR WORK.

Coach Painters' Green, Light.

carefully put away for future use. Remember that good work depends in a great measure on the strictest attention to cleanliness; and a sloven cannot in the nature of things produce a perfect job in Carriage Painting. "NEATNESS, ORDER AND ECONOMY," should be the motto in every paint shop.

The work which has been under way for a period of about five weeks may now be considered as finished. It may stand a few days to harden, and then be hung up. The bolts, etc., having been blacked off, and dry, the completed carriage should receive the first of repeated washings, which it is destined to undergo; but this clean, cold water washing should be done by an experienced hand; otherwise it is better left undone. If properly performed it will tend to harden the varnish, and will rather improve the general appearance. The finished vehicle now may be turned out for service, and there need be little apprehension that the painting will not prove a durable and creditable job. It might have been completed in much less time and have presented to the eye quite as good an appearance. A great many carriages are so finished, and they may, and do no doubt, stand the ordinary wear and tear of country roads pretty well; but for use on city pavements, *time* is an indispensable element, and it would not be safe to finish work for city wear in less than we have given to the job in hand, unless some other and shorter method be adopted.

For the last ten years ways and means have been devised, and every effort made to shorten the process of Carriage Painting; to expedite the work and turn it out in less time. The pace has not been fast enough for the "times;" and quicker,

[Continued on page 63.

Chrome Green.

This well known and serviceable pigment is another of the products of the metal chromium, in combination with iron and cyanogen (prussic acid). Chrome ore is a union of chromium and iron and is of rare occurrence. It is found in considerable quantities in Maryland, near Baltimore. The chief application of this ore is in the production of chromic acid (bi-chromate of potash), which is extensively used in dyeing and calico printing and in the production of the paints known as chrome colors. Chromic acid, literally color acid, derives its name (the Greek original signifies color, or to color) from the property it possesses of throwing down a colored precipitate when added to the saline solutions of certain metals.

The paint known as Chrome Green is made by mixing together Chromate of Lead (Chrome Yellow), and Prussiate of Iron (Prussian Blue). This product is combined with an earthy base, Sulphate of Baryta, or Silica, in proportions to suit the market as to price. It is sold under various names but the paints are the same, differing only in the quantity of coloring matter which they contain and the tones of color.

Coach Painters' (Superfine) Green, Dark. Price, per pound, in one pound cans, &c., 50 cents.

JOHN W. MASURY'S
Superfine Colors.

PREPARED ESPECIALLY FOR COACH AND CARRIAGE PAINTERS,
AND ORNAMENTAL AND CAR WORK.

Coach Painters' Green, Deep.

WILLIAM J. READ,
PRACTICAL
STEAM JOB PRINTER

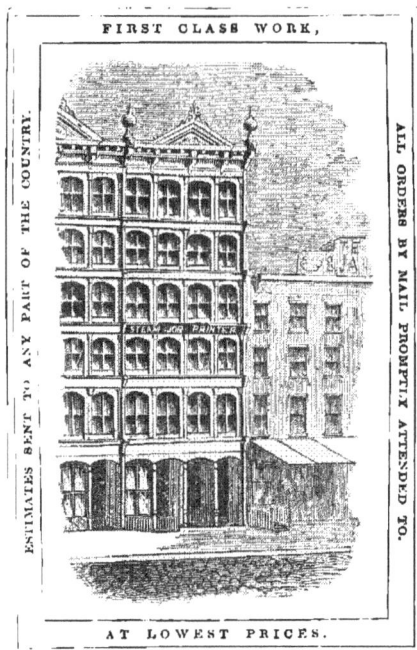

LITHOGRAPHING IN ALL ITS BRANCHES.

CARDS, CIRCULARS, BILL HEADS.

ESTIMATES SENT TO ANY PART OF THE COUNTRY.

FIRST CLASS WORK,

ALL ORDERS BY MAIL PROMPTLY ATTENDED TO.

PROGRAMMES, BALL & WEDDING CARDS.

BLANK BOOKS MADE TO ORDER.

AT LOWEST PRICES.

LAW CASES, POINTS & BLANKS.
116
FULTON STREET, N. Y.

shorter ways, of arriving at the same result have been sought for—if not discovered. Keeping in mind the grand, pervading principle of compensation, we are not of those who believe the time heretofore deemed necessary to produce a first-class job of Coach Painting can be materially shortened, at the same time retaining *all* the good features and results of the slow process. That is to say : the chances are altogether in favor of durability, when oil enough has been used in the painting to ensure elasticity and prevent the material from drying to that flinty hardness which cannot be supposed to bear the shaking and concussions which all wheeled vehicles on city pavements are necessarily subjected to, without cracking, and perhaps, chipping off. In short the mode of painting carriages such as we have described in the foregoing pages of this book, involves the expenditure of a certain number of days, which cannot be materially curtailed without incurring the risk of what has been too common of late, viz., jobs which soon perish with the using.

If haste be a *sine qua non* with the painter; if the work must be completed in half the time heretofore deemed essential in the production of enduring carriage painting, it is suggested that some other mode be adopted. If we will have railroad speed, we must abandon the stage-coach system !

In writing about the " new way," which certainly has found favor in many first-class manufactories, we propose simply to give the results of our own experience, without endorsing or committing ourselves either to the old system or the new, as possessing superior advantages. To give the facts as we find them is what we propose, leaving every man to his own judg-

[Continued on page 67

Naples Yellow.

This pigment is of no particular importance or interest to coach painters, It is said to be a compound of the metals Antimony, Lead and Zinc in unequal proportions, the first named metal being predominant.

It was at one time an important pigment, particularly in the fine arts, and its manufacture was confined to Naples, the mode of operation in its production being kept for many years a profound secret. Since the discovery and introduction of Chromate of Lead (Chrome Yellow), the demand for Naples Yellow has very much declined, the Chromate of Lead, being in all respects a superior paint, both in color, body, drying property and ease of working. Naples Yellow is a poor dryer, does not flat well, and great care is required to make solid work with it.

The sample shown on the opposite page is put on over a coat of ground made to match the color; and finished with one coat of clear color and one coat of color varnish.

Naples Yellow, L and D, ℔ ℔, $1 25.

JOHN W. MASURY'S
Superfine Colors.

PREPARED ESPECIALLY FOR COACH AND CARRIAGE PAINTERS,
AND ORNAMENTAL AND CAR WORK.

Naples Yellow, Light.

66

ment as to which course he will adopt or pursue. Something more than three years has, we believe, passed since the introduction of the "permanent wood-filling," and in the candid judgment of the writer, it has steadily grown in the favor of the Trade ; and the complaints about the cracking and chipping of paint from carriages are decreasing with the more general use of the new article. We have seen carriages painted over a priming of the wood-filling, in constant use for more than eighteen months on the city pavements, which did not show the least sign of cracking or chipping. True, some painters have complained that the work is more likely to perish over the wood-filling than when put on over the successive coats of lead, as in the old practice. If this misfortune did not sometimes occur under the old method, the objection would have more force. It is suggested that the rubbing varnish in such cases has more to answer for than the material used in the operation of priming. Any painter may do something toward settling this question by painting two panels, one with wood-filling underneath the roughstuff, and the other with lead and oil, and then treating the two processes alike up to the first coat of varnish. For convenience a moulding should be run through the center of each panel. On one-half of each use a quick-drying, hard, non-elastic and brittle varnish—there is no scarcity of such in the market—and on the other two moieties, use a varnish, which though drying hard, is not brittle, but tough and elastic ; a varnish which requires a good amount of labor to cut down. Finish all with a coat of best English varnish and abide the result. If those parts on which the brittle

[Continued on page 71.

Chrome Yellow Superfine, (Lemon)

This useful and brilliant Yellow results from a combination of chromic acid with lead in solution, and is properly speaking a chromate of lead. It has a dense body, particularly when ground to an impalpable fineness. It is much used in the paint shop, not only by itself, but also with white for making Pale Yellow, Buffs, Cream Color, etc.; and for producing with Black or Blue, the various broken Greens, as Olive Green and Quaker Green, and Bronze Green : and with Red, Orange color ; and with Red and Black, or Blue, the brighter hues of Brown. It is a permanent color when exposed to the light, but blackens when submitted to the fumes of sulphuretted hydrogen. A coating of varnish however, effectually prevents this.

Superfine C. P. Chrome Yellow (Lemon)

Price, ℔, in assorted cans, - - 50 cents.

Do not make a workshop of your varnish-room.

How to rub off a Varnish Run.—Wet your cloth and wipe it over a piece of hard soap. Use fine pumice stone as customary. The action of the soap will keep the particles of stone from sticking into the soft varnish.

JOHN W. MASURY'S
Superfine Colors.

PREPARED ESPECIALLY FOR COACH AND CARRIAGE PAINTERS,
AND ORNAMENTAL AND CAR WORK.

Chrome Yellow, Lemon.

substance was used do not perish within three months, while the other parts remain full and sound, we will cheerfully acknowledge the vanity of our own experiments in that direction. Taking this theory as the correct one, it follows that the work is liable to the accident in question without regard to whatever system may have been adopted in the initiatory proceedings, supposing a mistake in the selection of the varnish. Again : it is far better that a job should perish, than to crack and flake off ; because, in the one case the remedy may be found in re-varnishing, while the other involves the trouble and expense of burning off and re-painting. It is not to be supposed that any new claimant for public favor can find it, all at once. Many pertinaciously cling to what has been tried, and not found wanting. The bridge is good which carries safely! To those who would inquire further we recommend applying for information to any of the first-class establishments, where the use of the permanent wood-filling has been adopted. But to the *modus operandi.* Priming with filling should be proceeded with as in the use of lead. It must be put on evenly and well brushed into the grain of the wood, and under no circumstances must the beads and corners be left full of the material. A short, well-worn brush is best for applying it, and the work should stand four days before applying the first coat of roughstuff, which should be mixed as follows: two parts dry lead, one part English filling, wet with two parts Japan gold size, two parts varnish, and one part raw oil, mixed thick and ground finely through the mill. Reduce with turpentine and apply as usual. Putty on this coat after three days, and give the putty two days

[*Continued on page* 75.

Superfine C. P. Chrome Yellow, Deep Orange.

The remarks on the foregoing page are mainly applicable to this color, and there remains only to say that this very rich pure Orange makes of itself a good finish for carriage parts, and is much used for fancy Express Wagons.

Price, ℔ lb, in assorted cans - - - - 50 cents.

The two intermediate tones of this color are not shown by sample, mainly to avoid making the book too bulky. They are respectively darker than the Lemon, and considerably lighter than the Deep Orange.

Superfine Chrome Yellow, Orange.

Price, ℔ lb, in assorted cans, - - - - - - 50 cents.

Superfine Chrome Yellow, Medium.

Price, ℔ lb, assorted cans, - - - - - - - 50 cents.

Varnish should not be reduced with cold spirits of turpentine. If too stout for working with the brush, better return it to the maker.

JOHN W. MASURY'S
Superfine Colors.

PREPARED ESPECIALLY FOR COACH AND CARRIAGE PAINTERS,
AND ORNAMENTAL AND CAR WORK.

Chrome Yellow, Deep Orange.

to harden, before applying second coat. This should consist of two parts English filling mixed stiff in two parts gold size, one part raw oil and one part varnish. Thin with turpentine and allow two days for drying. Third coat should be as the second, excepting that the oil should be omitted and brown Japan substituted therefor. After one day, a fourth and last coat of roughstuff may be put on. This may be made, three parts English filling, one part dry lead, in two parts gold size, one part varnish and one part brown Japan. If a fifth coat be deemed necessary make it same as fourth coat. Apply guide coat, and rub and finish as in the old way. The carriage part, coming from the smiths should be trimmed up, bands put on, etc., and thoroughly sandpapered, cutting close down to the wood. Dust off carefully and apply coat of filling to every part, iron work included. Brush the wood-filling well into the grain, taking care not to use too much. A thin coat is best. Next day putty rims, faces, spokes, and all flat places evenly, with soft putty made elastic. The usual mode of proceeding is to smooth down next day for color ; but our practice has been to apply with a flat camel hair brush, a coat of lead, mixed stiff in Japan and reduced with turpentine to the consistency of color, previous to sandpapering. This will ensure a more perfect surface. This coat may be colored with a little black if the finish is to be Black, or with Indian Red, if the finish is to be Carmine or Lake. By adopting this mode of proceeding, the sandpaper will not be apt to clog, and tear up the filling, and if proper care be exhibited in rubbing down, the lead will come off, and there will remain a good surface without injury to

[Continued on page 79.

Yellow Lake.

The ground for this glazing is made of two parts of our Silver White (White Lead may be substituted for this), two parts Burnt Umber, one of our Medium C. P. Chrome Yellow, one of Yellow Lake, and a very little Scarlet Lake. This is a very useful, indeed indispensable, color in the paint shop, and the use of it gives a delicate look to various shades of drab, tan colors, greens, etc., which can be obtained in no other way. The rich, delicate drabs, which have been so long used by Eastern manufacturers, and which have been so much admired, are produced by a glazing of this Lake.

Yellow Lake A, in 1 ℔ and ½ ℔ cans, ℔, $2.00.
" " " " ¼ " " 2.25.
" " B, " " ½ " " 2.50.
" " " " ¼ " " 2.75.

Do not lose sight of the fact that all our ground colors are strictly pure. We never cheapen a color by adulterating it. We get the best of the kind, and offer it to you finer and in better shape for working than was ever offered before. Pure colors are with us a specialty.

JOHN W. MASURY'S
Superfine Colors.
PREPARED ESPECIALLY FOR COACH AND CARRIAGE PAINTERS,
AND ORNAMENTAL AND CAR WORK.

Yellow Lake, Glazing.

the foundation. Sandpaper the next day, dust off, and apply first coat of color made more elastic with oil and varnish than for coloring over lead paint. The second coat of color may be more elastic than the first. From this point all subsequent proceedings up to finish will be the same as in the old method.

Such has been our mode of proceeding in using the permanent wood-filling. There may be better and shorter methods, but the results of our experiments have been satisfactory.

Disclaiming any intention of dictating a rule of action for the conduct of others, we suggest a trial of the mode above described to those who have not given the matter any attention or trial. Every painter is supposed to have his own peculiar ways and notions as to how painting should be done. With these we have no desire or intention of interfering. The trade of Coach Painting is not to be classed with mere mechanic routine. It rises out of mechanical drudgery into the domain of art. The ability to perform such work in all its possible completeness and perfection is an accomplishment of which any man may be justly proud. It does not seem that any labor-saving machinery can be brought to bear upon it in such a way as to lessen the necessity for cunning and skill, for education and taste.

Referring again briefly to the new mode of carriage painting, we would remark that, the question of time, durability and cost being all involved in it, the subject is entitled to a careful investigation.

Having concluded this somewhat lengthened description of the two modes of procedure, in modern coach painting, we propose to record some of the reasons why coach painters should

[Continued on page 83.

Crimson Lake, B.

The ground for this color is made of three parts Superfine Ivory Drop Black, and one part of our Superfine Burnt Turkey Umber. This Lake has great strength and depth of color, and wonderful covering properties. It may be mixed with English Vermilion for common jobs and produce satisfactory results. Time of drying, when thinned wholly with turpentine, one hour.

A,	CRIMSON LAKE, in 1 and ½ ℔ cans, ℔ ℔,	- $1.50
A,	" " ¼ " "	- 1.75
B,	" " 1 and ½ " "	- 2.00
B,	" " ¼ " "	- 2.25

A piece of silk saturated with varnish, and then rinsed out with spirits of turpentine, and allowed to dry for half an hour, used as a wiper, will take all the fine pumice stone, lint and motes from a body preparatory to finishing. Go over every part carefully and gently.

One part rubbing varnish, two parts English varnish, and one part turpentine bottled up, well shaken and put away for a few days, will give you an easy running, quick dryer for striping. For striping Carmine add a little good Japan.

Crimson Lake, Deep Am.

adopt the use of our ground colors rather than depend upon the usual facilities of the paint-shop for producing them. We wish it understood, once and for all, that we never cheapen a color by mixing it with adulterating materials. We warrant them absolutely pure so far as our handling of them is concerned. Honesty in such matters we hold, not only to be the best policy, but what is of far higher import, the best principle. The extreme adulteration of paints, which has of late years become so great an evil as to work out its own cure, has not wholly grown out of a disposition on the part of the manufacturer to secure immoderate profits. The consumer has been most to blame, because of the ready credence he has given to the promises of needy and unscrupulous sellers, who have promised to give him more for his money than its worth. It would seem almost beyond belief that a coach painter would risk spoiling a job in the hope of saving a half-dollar on a gallon of varnish. Would such a case be a novelty? The adulteration of paints is so difficult of detection as to make the practice easy and comparatively safe. Take for example the article of Carmine. In a color so expensive as this a small percentage of adulteration makes a material reduction in the cost. A single ounce in a pound of this expensive color would afford a larger profit to the seller than is usually realized by those who sell it pure at first hands. Nor could this be detected in using by the most skillful and practised painter. The cheat would be revealed only by the untimely fading of the color, and that would be too late to remedy the evil. In carriage painting, immediate effects are less important than remoter consequences.

[Continued on page 87.

English Crimson Lake.

This splendid shade of Lake is obtained by glazing over a
ground made from four parts of our Tuscan Red to one part of
our best Ivory Drop Black. The sheet was prepared thus :
One coat of *ground*, one coat of clear Lake, thinned with turpentine, and a little good rubbing varnish, and one coat of color
varnish. This Lake looks well either on light or heavy work,
carriage parts or bodies, and is warranted to hold its color as
long as any in market.

English Crimson Lake A :

Price, ℔ ℔, in 1 ℔ and ½ ℔ cans,				- -	- $5.50.
" " "	¼	"		- -	- 5.75.
" ℔ oz.	"	1	"	- -	- .40.

English Crimson Lake B :

Price, ℔ ℔, in 1 ℔ and ½ ℔ cans,				- -	- $6.50.
" " "	¼	"		- -	- 6.75.
" ℔ oz.	"	2	"	- -	- .50.

FOR BURNING OFF OLD PAINT.—Procure ten or twelve feet
of rubber tubing (which comes of proper size for such pur-
poses), slip one end over the gas burner; put a burner on the
other end; light the gas, and hold the flame to the work. One
man can burn off more old paint with gas than two with hot
irons.

JOHN W. MASURY'S
Superfine Colors.

PREPARED ESPECIALLY FOR COACH AND CARRIAGE PAINTERS,
AND ORNAMENTAL AND CAR WORK.

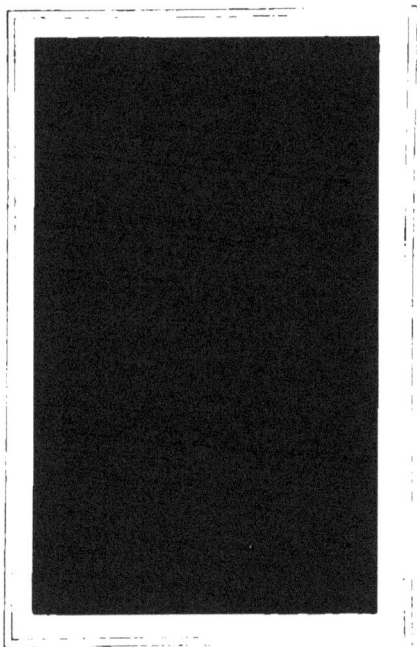

Crimson Lake, English Super.

FARRAR & ADAMS,

Carriage and Sleigh Manufacturers,

16 & 18 Portland Street.

Portland, Me., Feb. 11, 1871.

Mr. J. W. MASURY.

Dear Sir: We have used your goods for the last six months, and are perfectly satisfied with them in every respect, and cheerfully recommend them to Carriage Makers.

Yours truly,

FARRAR & ADAMS.

J. M. KIMBALL & CO.,

MANUFACTURERS OF

CARRIAGES AND SLEIGHS,

302 & 304 Congress Street.

Portland, Me., Feb. 11, 1871.

Mr. J. W. MASURY.

Sir: Having used your ground Colors for Carriage Painters' use for the last six months, we cheerfully say that they have given the most perfect satisfaction in every respect. Your Ivory Drop Black is indispensable, and the fancy Colors all that could be desired. We think that the Carriage Manufacturers throughout the country should unite in a testimonial to you for your efforts in their behalf.

Respectfully yours,

J. M. KIMBALL & CO.

86

The colors we offer are prepared expressly for use in carriage work, and with reference only to the requirements of the Trade. They are finer than it is possible to make them in the paint shop, for the reason that we make a specialty of this business, with means and appliances which outside of us do not exist. To illustrate the convenience of colors in the form we present them, suppose a case. Two or three new spokes in an old wheel are to be painted! The time necessary to prepare the paint from dry materials would be more than sufficient to match, paint, stripe and varnish with our colors ready at hand. Or, suppose an old carriage to be re-varnished. The color is mixed to match on the stone, and after, run through the mill. In the grinding process the color has changed, and is no longer a match. This may not be discovered until the application of varnish ; perhaps not even until the job is completed, and placed in a stronger light! The result is general dissatisfaction; but suppose it to have been discovered in the process of grinding! the change involves an addition of various colors; one after another is added, and with loss of much time, to say nothing of loss of patience, the result is a quantity of paint sufficient to paint two carriage parts, which of course is almost worthless for other work and finds its way into the waste, or slush-tub, as it is not very elegantly termed in the paint shop. Had our ground colors been on hand the match could have been made in one-quarter the time and with one-quarter the stock, and the saving would have been both in time and material, and the danger avoided of mismatching by the change of color in the process of grinding.

[*Continued on page* 91.

Carmine Lake, B.

The following sample is over a ground of four parts by measure of our best Ivory Drop Black, and one part of our Coach
Painters' Indian Red. Almost any desired shade can be obtained by using more or less of the latter color. It will dry,
ready for varnish, in from one to two hours, covers well, works
freely, and is well adapted for carriage parts or bodies. A little
wearing varnish is always good as a binder for all our fine
colors. Japan is not recommended for this purpose.

CARMINE LAKE,	B, in 1 and ½ ℔ cans, ⅌ ℔,			- -	$3.50
"	" B,	¼ " "		- -	3.75
"	" B,	2 oz. "	oz.	- -	25
"	" A, in 1 and ½ ℔ cans, ⅌ ℔,			- -	2.50
"	" A,	¼ " "		- -	2.75
"	" A,	2 oz.,	oz.	- -	20

N. B.—After your cans are open, care should be taken to
keep the color well scraped down, and the top covered with turpentine, and the can covered as closely as may be. Observe
this, and the result will be a saving of money, and good clean
working color.

[There is now no law which prevents painters keeping their
hands and clothes clean.]

JOHN W. MASURY'S
Superfine Colors.

PREPARED ESPECIALLY FOR COACH AND CARRIAGE PAINTERS,
AND ORNAMENTAL AND CAR WORK.

Carmine Lake.

DUSENBURY & VAN DUSER,

LIGHT WAGON MANUFACTURERS,

135 & 137 Christie Street,

NEW YORK.

J. W. MASURY, Esq.,

Dear Sir: Your Superfine Ground Colors have been used in our establishment since their introduction about a year since, and we take pleasure in stating that they have given perfect satisfaction in every respect, saving the labor and waste of grinding, and being of a fineness which we could not attain.

We would recommend them to the trade on score of convenience and economy, and wishing you success in your enterprise,

We are very truly yours,

DUSENBURY & VAN DUSER.

R. M. STIVERS,

FIRST CLASS CARRIAGES AND WAGONS,

FOR THE PARK, ROAD AND TRACK,

144 to 152 East 31st Street,

Between Third and Fourth Avenues.

New York, Feb. 13, 1871.

Mr. J. W. MASURY,

Sir: I have used your Fine Colors for considerable time past and find them perfectly satisfactory.

Yours, &c.,

R. M. STIVERS.

It may be said that a thorough-going practical painter does not make such mistakes; but such work is not always done by that style of workmen. It is often entrusted to boys and other persons of immature judgments; and in spite of all that may be said, such mistakes do happen in the best-regulated shops. Suppose another case. A new body is ready for color; an ordered job; promised on a certain day. Time is limited, and a mistake now is little less than a crime. The paint shop is short of hands! The foreman, driven with other work, finds just time to mix the black on the stone—after the same has been powdered by the boy—put it into the mill, turn the screw, and give pressure enough to insure moderately fine color. The day is a hot one—the crank turns slowly under the best efforts of the perspiring juvenile, who, like Mantilini, feels his life to be "one demnition grind;" tired and disgusted—not appreciating the importance of fine colors,—he gives the thumb-screw a half turn, and presto! the crank goes to a lively tune, the color comes out in no stinted quantity, and soon the task is at an end. Leaving the mill, which he neglects to clean, and the pot of half-ground color, and feeling himself entitled to a half hour's recreation, in reward for his industry and perseverance, he disappears, and the foreman comes from the varnish-room, with just enough of daylight left to color the body. The application of a single brush full of the paint informs him that in fineness it is equal to No. 2 sandpaper; but there is no time to grind a fresh lot! and the cup of thinned color could not be made fine in a week; so the boy, being found, is presented with a coat of—well, not blessings!—the body, unpainted, stands till next day, or, being

[Continued on page 95.

Munich Lake, B.

Over a ground work made of equal parts of our Superfine
Ivory Drop Black and Coach Painters' Indian Red.

This Lake is, in some parts, more in favor than any other, and
is a general favorite. As a rule, it holds its color well. It will
dry in about two hours, but we recommend, if time will permit,
the use of a small quantity of varnish and raw oil. Reduce
with turpentine. Bear in mind that we disclaim all part or lot
in the short-comings of unskillful varnish makers. No paint is
good, of course, with poor varnish. The best of its kind is
always the cheapest.

Munich Lake, B, in 1 and ½ ℔ cans, ℔ lb.	-	$4.50
"	"	B, "	¼	"	"	- 4.75
"	"	B, "	2 oz.	"	oz.	- 35
"	"	A, in 1 and ½ ℔ cans, ℔ lb.	-	3.25
"	"	A, "	¼	"	"	- 3.50
"	"	A, "	2 oz.	"	oz.	- 30

N. B.—That all our Lakes, Carmines and Blacks are ground
for drying quickly. The drying can always be retarded to suit,
by the use of a little oil and varnish.

JOHN W. MASURY'S
Superfine Colors.
PREPARED ESPECIALLY FOR COACH AND CARRIAGE PAINTERS, AND ORNAMENTAL AND CAR WORK.

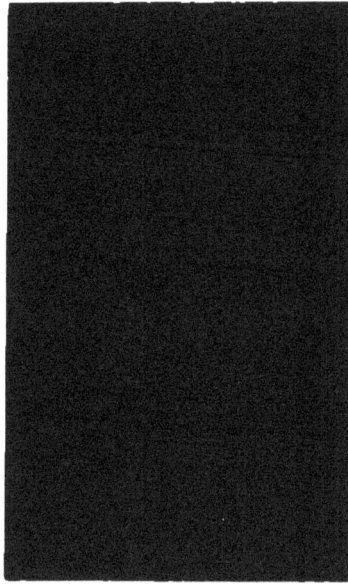

Munich Lake.

smalted, the surface requires an extra coat of rubbing varnish to present a respectable appearance. Do not such accidents frequently occur in the paint shop? Our ground colors offer a remedy sure, safe and economical for all these complaints! Try them and be convinced.

Looking at the question in an economical view, we give the following facts : Taking our Munich Lake, for example: seventy-three sheets, in size 19x29 inches, were coated with one pound of ground Munich Lake B. This gives a superficial area of say two hundred and eighty square feet, one coat clear color and one of color varnish, painted with a single pound of Lake. Let the sample be placed in sunlight, and there will be no difficulty in determining whether or not the work is solid. This would seem to settle the question and leave no room to doubt that our superfine colors are just what are required for first-class—or for any class work.

Our Mr. Wolcott says : These colors should come into general use, not only because they are *finer* than any other colors, but because they work more freely, flat more perfectly, and dry more readily than any others. This judgment he renders after having painted two thousand four hundred sheets —the quantity required for our first edition of five thousand copies of this work—equal to an area of about nine thousand two hundred square feet. Mr. Wolcott further says : Notwithstanding the fact that the spongy, porous nature of the paper renders painting the same much more difficult than over the hard, smooth surface of a panel, we can go over a sheet of the size named above, brushing the color four different ways

[*Continued on page* 99.

English Purple Lake, Deep.

This sample is worked over a ground made five parts of our Superfine Ivory Drop Black to one part of our Coach Painters' Indian Red. (As a hint, we suggest the substitution of two or three parts of our Superfine Ultramarine Blue for the Black. That is, say two parts Blue, two parts Black, and one part Indian Red or Tuscan Red.)

This Lake is considered by many as the most desirable of all. It is much used on bodies, and is generally supposed to be a more durable Lake than any other. Its use would have been much more general but for the difficulty in reducing it to perfect fineness. The use of a single can of our English Purple, will convince painters that it is not only ground fine, but that it works well and looks well.

ENG. PURPLE LAKE (Sup.), in 1 and ½ ℔ cans, ℗ ℔, - $7.00
"   "   "   "   "   ¼   "   "   - 7.25
"   "   "   "   "   2 oz. " ℗ oz.   - ' 50
"   "   " B, deep & ex. deep, in 1 & ½ ℔ cans,   5.00
"   "   " B,   "   "   "   ¼   "   5.25
"   "   " B,   "   "   "   2 oz.   "   40

N. B.—Painters should never use the chamois skin used on carriage work for a towel.

JOHN W. MASURY'S
Superfine Colors.

PREPARED ESPECIALLY FOR COACH AND CARRIAGE PAINTERS, AND ORNAMENTAL AND CAR WORK.

Purple Lake.

before it sets, and yet we can varnish over it in from one-half an hour to two hours. After conversing with more than five hundred painters as to the cost of grinding colors in the shop —the extremes in the estimates given being thirty cents as the minimum and one dollar as the maximum average cost for labor alone, and a waste of from ten to fifteen per cent.—we think we may aver, that our prepared colors, on the score of economy alone—to say nothing of all the other advantages— are worthy the attention of all who buy and use paint.

The " Pitting " of Varnish.

There is somewhere a proverb which runneth thus, or words to like effect : "To cease to justify one's deeds unto one's self is the last infirmity of evil." Coach painters, as a rule, are very wise, very learned, and have reasons to account for all ordinary and extraordinary phenomena, as "plenty as blackberries;" but I ask, in all sincerity, did anybody ever hear one—or see anybody who ever did hear one—account for a spoiled job by charging it to his own carelessness or neglect? Not that members of our craft are singular in this respect! All men are disposed to "justify themselves;" or, in other words, no man will load his own shoulders with blame, if he can with any show of reason, shift it so that another back shall bear the burden. A dozen painters will at a word give a reason for the pitting of Varnish. But as no two will perhaps agree, each of the statements must be taken with a liberal allowance of salt. A case in point: a foreman in a first-class city shop was using

[Continued on page 103.

English Scarlet Lake, C.

Painted over a ground of Deep English Vermilion. When a brilliant Scarlet Red is required this Lake has no superior. It has great body and strength of color. The shade of scarlet may be deepened by making the ground-work darker. Scarlet Lake B, not shown in the book is lighter than A, and not so light as C.

We have three qualities, which are distinguished by the letters A, B and C respectively, the last letter designating the best or highest price. These colors are ground the same as the Carmine and will dry, thinned with clear turpentine, in about half an hour.

SCARLET LAKE, ENG. C, ℔ in 1 ℔ cans,						-	-	$6 50
"	"	"	"	½	"	-	-	6 50
"	"	"	"	¼	"	-	-	6 75
"	"	"	℔ oz.	2 oz.	"	-	-	50
"	"	B, ℔ in 1 ℔	"			-	-	5 50
"	"	"	"	½	"	-	-	5 50
"	"	"	"	¼	"	-	-	5 75
"	"	"	℔ oz.	2 oz.		-	-	40
"	"	A, ℔ in 1 ℔				-	-	4 50
"	"	"	"	½	"	-	-	4 50
"	"	"	"	¼	"	-	-	4 75
"	"	"	℔ oz.	2 oz.		-	-	35

JOHN W. MASURY'S
Superfine Colors.

PREPARED ESPECIALLY FOR COACH AND CARRIAGE PAINTERS,
AND ORNAMENTAL AND CAR WORK.

Scarlet Lake, Light.

JOHN W. PERKINS & CO.,

PORTLAND, MAINE,

Wholesale Drug and Paint Dealer,

PATENT MEDICINES,

AND FINE

Chemicals, Paints, Oils, Varnishes,

AND GROCERS' DRUGS.

ALSO, A FULL ASSORTMENT OF

JOHN W. MASURY'S READY-MADE COLORS,

For all kinds of House Painting,

AND OF HIS

Superfine Colors for Coach, Carriage and Sleigh Painting.

THESE READY-MADE COLORS have been in use for years, and have in every case given entire satisfaction. Buyers are cautioned against imitations, and requested to see that every package bears our full name, "MASURY & WHITON, NEW YORK," and our copyrighted title, "RAIL ROAD COLORS.

A FULL ASSORTMENT OF COLORS, DRY AND IN OIL, ALWAYS ON HAND.

our ground colors. A hundred jobs had been turned out painted with these colors, which were in every respect satisfactory.
One day the first coat of Varnish did not flow smoothly. The
cause? Oh! those ground colors, of course! Send them back!
Don't use any more! Now, gentle reader, this thing had oc-
curred in that shop many, many times before, but then there
were no ground colors to make a scape-goat of. "To cease to
justify one's deeds unto one's self is the last infirmity of evil."

A story told many years ago in the *Knickerbocker Magazine*
may not be out of place here. It was of an old, ugly, ill-tem-
pered, cross-grained country village loafer, who was always
doing some ill-natured thing. It came to the ears of a fond
pater familias that a pet lamb, the object of his pet children's
affections, had been kicked by this aforesaid ugly customer;
and, full of indignation, the aggrieved father sought out the
offender and demanded why and wherefore this assault had
been made on the unoffending "pet." Ready with a reason,
the old curmudgeon replied in this wise: "I'll tell why I did it!
*That lamb tried to bite me, and I'll kick any cussed lamb that tries
to bite me!*" "To cease," &c., but we will not repeat the text
again. Now, in making the application of this little story, we
expect to be made the scape-goat of many sins; but we do not
like to be accused of trying "to bite" anybody. If Varnish,
previous to the introduction of our Superfine Colors, had not
been known to have presented a "pitted" surface, it would be fair
to ascribe to said Colors the unfortunate result; but, as this
thing has been known as long as Varnish has been used, it would
be reasonable, at least, to look behind our Colors for the cause.

[*Continued on page* 107.

Scarlet Lake, B.

The sample of Scarlet Lake is painted over a ground of our Coach Painters' Indian Red. One coat of Lake and one coat of color varnish. The color has a strong body and covers well. Thinned with clear turpentine, it dries in from one to two hours. If you are not pressed for time, a little wearing body-varnish, as a binder, may be used with good results, and a very little raw oil. The latter, however, must be used with great care, as Scarlet Lake is a non-dryer to the last degree. Otherwise this color is to be handled in all respects as Carmine.

SCARLET LAKE, C, in 1 and ½ ℔ cans, ℔, - $6.50
"   " C,   ¼   "   "   - 6.75
"   " C,   2 oz. "   oz. - 50
"   " B, in 1 and ½ ℔ cans, ℔, - 5.50
"   " B,   ¼   "   "   - 5.75
"   " B,   2 oz. "   oz. - 40
"   " A, in 1 and ½ ℔ cans, ℔, - 4.50
"   " A,   ¼   "   "   - 4.75
"   " A,   2 oz. "   "   - 35

N. B.—A continued washing of one's hands in spirits of turpentine will almost certainly result in stiff joints.

JOHN W. MASURY'S
Superfine Colors.

PREPARED ESPECIALLY FOR COACH AND CARRIAGE PAINTERS,
AND ORNAMENTAL AND CAR WORK.

Scarlet Lake, Deep.

Will those who are so ready to find a place whereon to rest the blame of a spoiled job, bear in mind the fact that all our Coach Painters' Colors are prepared under the immediate supervision of an intelligent man who has had an experience of more than a quarter of a century in the paint shop! It is well to be wise, but not good to be wise in one's own conceit! There are to-day hundreds of painters throughout the country using our Superfine Colors. In one or two solitary instances a complaint comes that the Varnish "pits" over our Black. Now, as this does not happen in the ninety and nine cases, but only in the one hundredth, we claim that it is only fair to look for some other cause before condemning the Color. It must be kept in mind that Coach Painting is an art, and that something more is required to perform it satisfactorily than just enough of knowledge to mix a cup of color and apply it. Good judgment, sound discretion, close observation, no less than a cunning hand, are the *sine qua non* of success. Every painter, whose vision is not limited by the end of his nose, is aware that Varnish is apt to "pit" on a Japan gloss; especially if the color be not quite hard. Our Colors are mixed and ground with a purpose to avoid the use of Japan, in thinning for application. Turpentine for thinning, with a little hard-drying, rubbing varnish for a binder (and a drop or two of oil, if you please), are all that is required. If one *will* use Japan in place of varnish, we beg of him to use something which is entitled to be called Japan, and not the stuff which is hawked around the country under that name, but which would almost thicken turpentine when mixed with it.

[*Continued on page* 111.

Carmine Pure, No. 40.

Pure French Carmine, of brilliant color and unsurpassed strength. The sample on the opposite page is put on over a ground of our deep English Vermilion, one coat of clear Carmine, thinned with turpentine, and one coat of color varnish. The Carmine coat was dry enough for the color varnish in one hour ; but we do not advise working it so quickly, except in cases of actual necessity. If not driven, we advise the use of a small quantity of raw oil, which gives more time to work, and, therefore, ensures a more perfect solidity. We have found a first-rate wearing body-varnish best for glazing Carmine jobs. The use of a small quantity of our Indian Red or Tuscan Red with the Vermillion, for ground-work, makes it much easier to cover with Carmine.

For this color, as for all other transparent colors, use no other brush than that known among painters as flat camel's hair, but to the brush trade as *spalters.*

Bear in mind the fact that all our colors are perfectly pure and of the best quality.

PRICE OF CARMINE IN 1 AND ½ ℔. CANS, - - $11.00
" " ¼ " - - 11.50
" " PER OUNCE, 2 OZ., - - 75

JOHN W. MASURY'S
Superfine Colors.

PREPARED ESPECIALLY FOR COACH AND CARRIAGE PAINTERS,
AND ORNAMENTAL AND CAR WORK.

Pure No. 40 Carmine, Deep.

BAILEY & YORK,

Manufacturers of Carriages and Sleighs,

Nos. 41, 43 & 45 BOWKER STREET,

Boston, Feb'y 2, 1871.

I have used the "Ground Paints," manufactured by John W. Masury, of New York, and shall continue to do so as long as I can get them—that I consider them the finest, purest, and best material ever used in a paint shop.

FRED. J. GREENE, Painter,
With BAILEY & YORK.

Boston, Feb'y 3, 1871.

J. W. MASURY, New York.

Dear Sir:—Some eight months ago we adopted the use of your "Superfine Colors," and the frequency of our orders must convince you of the high estimation in which we hold them. They never come short of all that is claimed for them, and with us they have been a perfect success.   Very truly yours,

D. P. NICHOLS & CO.

Boston, Feb'y 3, 1871.

J. W. MASURY, New York.

Dear Sir:—Your "Superfine Ground Colors" have been used in my shop for some six months, and have given the best of satisfaction to myself, and also to my painters.

Very respectfully,   CHAUNCY THOMAS.

I cheerfully indorse the above.

LEONARD B. NICHOLS,
Foreman Painter.

110

Let us enforce, by a few words, a wise injunction, the result of inspired wisdom. " Be not wise above what is written." Every man is supposed to be the best authority as to how to mix and thin colors which he has himself manufactured from the beginning. We rightly, we think, claim the same authority respecting the Colors which have been mixed and ground by us. We have put all these colors to actual test—witness, our Sample Cards—and we write what we have found out by actual experience. In the painting of four thousand sheets, including all the Colors, there was not a single instance where the varnish did not flow smoothly. So, we think we have the right to suggest that the best results may be produced by following our written suggestion, rather than by being "wise above what is written."

Trials and Difficulties.

The introduction of a new article—or an old one in a new form—is under the most favorable circumstances attended with many difficulties. Not only are there the prejudices even of the skillful and honest workman to contend with, but the besotted prejudices of the ignorant and stubborn; as, also, the "tricks of the trade," which unscrupulous salesmen know too well how to use to give a bad name to goods with which they cannot successfully compete in honest and fair competition; to say nothing of "the ways that are dark" among the very men of all others who would be supposed to encourage any improvement which would

[*Continued on page* 115.

Pure No. 40 Carmine, Light.

This color is put on over a groundwork of JOHN W.
MASURY'S ENGLISH, PURE, DARK VERMILION. One coat of the
Carmine, thinned with clear turpentine, and one coat of Color
Varnish.

This Carmine, when thinned with clear turpentine, will
dry ready for varnish in one hour in a warm room—that is, in
a room heated to about seventy degrees Fahrenheit. If there
be no great hurry for the job, the use of a *small* quantity of raw
oil is recommended, as the same will make the color work more
freely, and, by giving more time, does not drive the workman
to such an extent. A first quality wearing body varnish is best
for glazing Carmine jobs. The addition of a proportion of our
"Coach Painters' Indian Red"—say, one part of Red to two of
Vermilion, for the groundwork—will make it easier for the
painter to produce a solid, well-covered job.

For applying this, as for all transparent colors, use nothing
but what are known among painters as flat camel-hair brushes.
These are known in the brush trade as camel-hair spalters.

JOHN W. MASURY'S
Superfine Colors.

PREPARED ESPECIALLY FOR COACH AND CARRIAGE PAINTERS,
AND ORNAMENTAL AND CAR WORK.

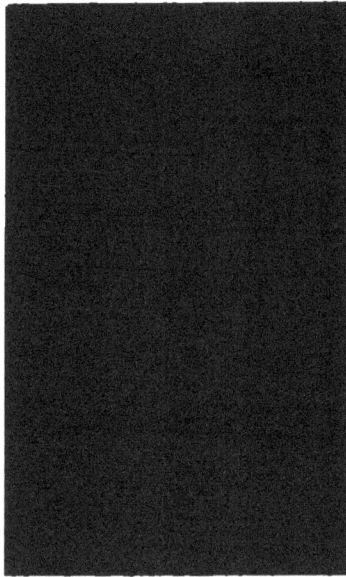

Pure No. 40 Carmine, Light.

OLD HOUSE OF BREWSTER.

ESTABLISHED 1838.

65 EAST 25TH STREET,

New York.

OUR SPECIALTIES:

100 lb. Wagons, Demi-Landau, Derby Wagons, 3 styles.

ALL THE ORDINARY STYLES CONSTANTLY ON HAND.

All Wagons are made with our Vertical Steel Plates, combining the Greatest Strength with the Least Weight.

65 *East 25th St., New York, Feb'y* 8, 1871.

J. W. MASURY, Esq., 111 Fulton St., N. Y.

Dear Sir:—We have used your "Superfine Ground Colors" for some time past, and can cheerfully recommend them as first-class in every particular.   Very respectfully yours,

J. B. BREWSTER & CO.,
65 East 25th St.

Bridgeport, May 1, 1870.

This is to certify, that, having used Mr. John W. Masury's (successor to Masury & Whiton) "Fine Ground Colors," I find them superior to anything ever used, in every respect.

P. MUNDRY,
Ornamental Painter of Housatonic R. R. Co.

114

seem to lessen the drudgery of their daily occupations. Illustrative of the last named difficulty, let us give a case in point. The foreman of one of our city shops after using our Ground Colors for months with entire satisfaction and written commendation, all at once discovered that he could not use them more, because the varnish "pitted" over the black; which certainly were "pity, if 'twere true." Thinking to overcome the difficulty—not at first seeing the "cullered pussun in the fence" —we called upon the disaffected one, and proposed to try the Black then and there, and wait the drying and see with our own eyes this most wonderful phenomenon. Curious to relate, there was nothing just then which could, by any possibility, be used to make the test. *Mirabile dictu!* Not a gig-lamp even! Not a spoke, old or new, which would bear a coat of Black; and the *innocent* could not even hazard a guess as to when there would probably be any work ready for a coat of Black. As a last resort, we begged a small vial full of this varnish which behaved in such a *pitiful* manner, *only* when covering a coat of our Superfine Ivory Jet Black. Thankful for this, we went home and proceeded to coat, with some of the identical Color, four spokes; which were finished in black and varnished more than a year before. This was completed just at night-fall, and the next morning we applied to spoke number one, which presented a surface smooth as enamel, a coat of the aforesaid varnish. To spoke number two, we applied a coat of the same varnish, mixed with the color. To number three we gave a coat of the same varnish mixed with two other varnishes, one of which was a hard-polishing, and the other a very elastic

[*Continued on page* 119.]

Masury's New Green.

This pure, brilliant and permanent light green seems to supply a want which has long existed, viz.: a green which should possess the durability and non-fading properties of the so-called Paris or Emerald Green—with the brilliance and pure tone which no other green gives—and at the same time admit of being ground to that perfect fineness indispensable to the best carriage work, without losing color. In the belief that our New Green does possess all these good qualities, we offer it as a not expensive substitute for all the light greens used in coach painting. The sample on opposite page was put on over a ground of our Coach Painter's Green, light, mixed with enough white to produce a tone of color most like the pure green possible. Two coats of pure green, mixed mostly with turpentine (a little good rubbing varnish added to make the color flow well) were put on over the coat of body green. This green works as well in the brush as any color, and dries for varnish in a few hours.

In assorted cans, one pound and upward, ℔ ℔, 75 cts.

JOHN W. MASURY'S
Superfine Colors.

PREPARED ESPECIALLY FOR COACH AND CARRIAGE PAINTERS,
AND ORNAMENTAL AND CAR WORK.

Masury's New Green.

Mr. J. W. MASURY.
Boston, Feb'y 1st, 1871.

Dear Sir :—Having sold your "Fine Colors" for the past three months, we find the demand for them constantly increasing, and we are happy to add that wherever they have been used they have given perfect satisfaction.   Yours, very truly,

WADSWORTH, HOWLAND & CO.

Messrs. MASURY & WHITON.
Buffalo, Sept. 19, 1870.

Gentlemen :—We take pleasure in stating that we have been using your "Superfine Ground Colors" for the past six months, and find them *all* they are represented to be.   They are ground very much finer than we can grind them, work easy, flat and dry well; and we find them to be much cheaper than dry colors, waste and labor of grinding added.   We most cheerfully recommend them to all carriage makers, as cheap, reliable, and just what is wanted.

Yours, truly,   HARVEY & WALLACE.
J. C. SNYDER.

J. W. MASURY, New York.
Amesbury, Mass., Feb'y 7, 1871.

Dear Sir :—We, the undersigned, carriage manufacturers of Amesbury, after using your "Superfine Ground Colors" for a sufficient length of time to determine as to their merits, do *now* freely testify that we consider them unequalled for fineness, drying, and working qualities, and believe them to be the most economical paints ever put upon the market.   Wishing you the success which your efforts so richly deserve,

We remain very truly yours,

E. S. FELCH,
CHESWELL & BOARDMAN,
C. W. PATTEN,
HUME & MORRILL,
R. F. BRIGGS,
F. D. PARRY,
J. R. HUNTINGTON,
S. & C. ROWELL,
CARR & ALLEN,
A. M. WATERHOUSE, West Amesbury.

varnish. With number four we made another most severe test. In every case the result was a surface as smooth—aye, smoother than the most highly polished plate glass. "Was it not pitiful?" We then concluded that Ah Sin had "been seen," as the politicians put it; or, that he wished to be "at the game he did not understand!"

How to make the best Job in Black.

A pure Black is, in theory, the absence of all the primary colors and of the extreme color, White. The presence of any one of these detracts from the entireness of black. So, when black is viewed through any colored medium it ceases to be pure black, and assumes that tone of color which would result from *mixing* the color of the medium with the black. For example: Black, when viewed through a medium of yellowish varnish, reflects, however slightly, a greenish hue; and the greater the number of coats of clear varnish, the greater will be the detraction from the purity of the black! So with white! A single thin coat of the palest varnish over a coat of pure white detracts slightly from its purity. But successive coats, of the most colorless varnish, destroy the whiteness, and the surface reflects more or less of impure yellowish light. The same may be said of all the primary and secondary colors. Some of the mixed and broken colors would be improved, on the contrary, by a coating of a yellowish translucent medium; as, yellow lake over drab or over a mixed green. (See colored

[*Continued on page* 122.

Deep Green.

The sample on the opposite page shows one of the colors which can be produced only in the paint shop, and in the progress of the work of painting. Any tone of this color may be produced by mixing together Prussian Blue and Yellow Lake, and adding a small quantity of Lemon Chrome Yellow. The ground should be prepared thus : With Prussian Blue and Chrome Yellow make the ground as near the tone of color you desire to finish as possible. Put into the first coat of rubbing varnish a little Yellow Lake—say, two spoonfulls of Lake to a cup of varnish. In thus proceeding you get a green of depth and richness which can be obtained in no other way To satisfy any of this, let the sample be held in the sunshine.

In case any one of our ground colors should be found too thick to incorporate readily with the thinning, it may be run loosely through the mill so as to mix it perfectly. A cupful may be run through in a few minutes. We purpose to have *all* our colors of such a consistency as to admit of thinning readily, but mistakes will sometimes occur.

JOHN W. MASURY'S
Superfine Colors.

PREPARED ESPECIALLY FOR COACH AND CARRIAGE PAINTERS,
AND ORNAMENTAL AND CAR WORK.

Very Deep Green.

examples on pages 77 and 123.) In avoidance of these accidents, and in order to secure the best results possible in carriage painting, we suggest the application of only one coat of clear varnish, and that, of course, the last one. We believe the best work turned out of any city establishment is finished without a single coat of clear color (we speak now more particularly of glazing jobs), and with but one coat of clear varnish. In carmine and the lakes, the first coat on the ground is put on in varnish, and every coat of varnish up to the last is colored. In this way a depth of color is obtained which can be had by no other process. It should be borne in mind that the opaque, or body colors do not compare in beauty and brilliancy with the transparent colors! And, as a rule, the colors are beautiful in proportion as they are transparent. For examples : ultramarine blue, carmine, emerald green, scarlet and crimson lakes, &c. All are familiar with the beautiful colors reflected from the vases placed in the windows of apothecary shops. This results from the *depth* of colored fluid. A thin, flat glass vessel would not reflect such hues, though filled with the same substance. The principle is the same in carriage painting! To show the best possible colors, the light must be reflected, not from a flat, opaque surface, but from a surface which has beneath it, a depth of continuous colored particles reaching away down through the successive coats of varnish to the groundwork. To be sure, this mode of proceeding is expensive, both in labor and material! but who ever gained any good thing without working for it? Black should be put on one coat of clear flat color; after that, every coat of varnish should contain more or less of the same black as used for the first coat, up to the finishing coat, which should be clear varnish. In this mode the black holds its color, and does not take on the greenish tinge, which otherwise it is impossible to avoid. All work, of course, is good or bad only by comparison. Any carriage is black enough in a dark night! and almost any tolerably good black looks well enough when viewed *per se.* It is only when placed in comparison with the best, that its inferiority is apparent; and men who strive to excel in their productions are not content to occupy inferior positions in any particular. "Excelsior" is a good motto for coach painters !

Waste of Paints through Negligence and Ignorance.

The money value of paint wasted in this country is enormous. Greater, perhaps, than in all the world beside. Our reckless prodigality, in a certain way, is only equalled by our absurd attempts at economy. For example : A painter will sometimes spend the time and exertion necessary to walk a mile, all for the purpose of purchasing a can of paint a shilling less than he can buy it for under his very nose; and then neglect the proper precaution and preventive to waste, by omitting to cover up and take care of whatever paint may be remaining after the job is finished. Now, a quarter part of the time and labor necessarily expended in saving the shilling, devoted to care and cleanliness, would have resulted in the saving of twice that amount. Another absurdity! a slavish devotion to names. When will men learn that two things are not *necessarily* the same, because they may be called by similar names? Take, for example, the greens used in carriage painting. These are either chrome or copper greens, and are briefly described in the foregoing pages. The *body* greens are chrome colors, and the diaphanous greens are copper colors, as a rule. Chrome Green, when pure, is of a dense body (almost, in this respect, rivaling lampblack), and covers and conceals all it touches, whether white or black. A fair selling price of this pure green, dry, would be about eighty cents per pound to consumers, and at this price it would be the cheapest green attainable. Yet the probabilities are, that a pound of this color, under its own proper name, cannot be found in any carriage shop in the United States. What, then, do we buy? Listen! The so-called "chrome green" of commerce is simply an earthy base—silica, sulphate of baryta, or carbonate of lime, colored with chrome green, in proportions varying from (the best), say, one pound of color to five pounds of the base, to one pound of color and two hundred pounds of the base; and all is sold as chrome green. Now, this earthy base, which is transparent when mixed in oils, adds to the value of paint in the same manner and degree as watering milk, sanding sugar, or mixing shoddy with wool in the production of cloth adds to the value of these articles respectively. What the painter requires is *color*, *not* sand! And, considering that he has to pay vastly more for the color he buys when mixed *with* the sand, it would seem not to require a very elaborate argument to convince the dullest com-

[*Continued on page* 126.

JOHN W. MASURY'S
Superfine Colors.

PREPARED ESPECIALLY FOR COACH AND CARRIAGE PAINTERS,
AND ORNAMENTAL AND CAR WORK.

Car Body Color, Extra.

prehension that, for the consumer, *pure* colors are the cheapest. Some idea of the coloring property or power of real chrome green may be had by reflecting on the fact, that a single pound of it will impart its tone of color to a hundred pounds of a glassy translucent substance, causing it, in the mass, to resemble the pure green itself. I say "in the mass," for when this pretended green paint is spread upon a piece of glass and viewed through a microscope, or magnifying glass, it presents the appearance of vitrous minute grains, with a speck of color here and there, like small sea-birds scattered along a sandy beach. That a painter had better buy the color unmixed with the sand would seem a self-evident proposition.

We looked into a paint shop not long ago and espied a can, with our label on it, signifying that it (the can) did contain, or had contained, ivory black in Japan. It was uncovered and exposed to the dust, dirt and drying influences of a warm shop. We looked into it and found it contained about two-thirds of its original contents; but of its original value not a fifteenth part was there. One after another of the hands, in want of a little black, had dipped into it with palette knife, and the deep pits, or holes, were left, unfilled, to dry around the sides, and thus waste the material in the speediest possible manner. We asked the foreman how he was suited with our goods. "Oh!" said he, "the black is first rate, but it *dries up so!*" We thought, if it did not dry under such treatment as that, it would well deserve any amount of maledictions.

The utmost care and attention and the most scrupulous cleanliness are indispensable to economy and good results in the CARRIAGE PAINT SHOP.

Ivory Surface Enamel White.

The introduction of our New White leads to a new system of painting, the inauguration of which may not prove of special benefit to those of the Trade whose apprenticeship to the business has been in propelling the ladder-cart or turning the crank. The new system does certainly require skill, judgment and taste. Therefore let no "dauber" attempt it, because his failure will be complete. 'Tis true, a plain white flat job can be as easily executed with our New White as in the ordinary way; as can also a job of gloss work. Indeed, for gloss work, it is as much ahead of the zinc white and Demar gloss as best English coach varnish is superior to rosin varnish; but, to finish a job which shall show out in full perfection all that the new system is capable of, requires a skillful and finished workman. The objection we anticipate, on the part of painters, is, that a job once done in the new way will not require renewing short of twenty years. That objection, however, if it be an objection, is inherent in the system, and inseparable from it.

Gloss finish is not objectionable *per se*, and would not have fallen into disfavor, only that there was no way of doing it so as to make a job in any wise perfect. The ordinary mode of finishing such work—that is, by mixing a portion of zinc in Demar varnish—was not attended with satisfactory results. The varnish always dried with a tack, which remained for a long time; and when it hardened so as not to be sticky, it was liable to crack in all directions. The difficulty of making the work look solid, the disagreeable working quality of Demar, its liability to run, and the cheap-looking gloss which the surface presents, all have served to render it unpopular, not only among the Trade, but with householders generally. The New White has entirely different qualities, and works and finishes altogether differently. It makes solid work, dries as free from tack as French porcelain, yet is elastic and wearing to the last degree. It is suitable for outside work—as stages, wagons, cars, etc. The finishing or varnish coat does not require any preparation of material, the same being of a proper consistency to apply, without thinning or other manipulation, except, when the package has been for a long time at rest, to stir or shake until the whole is of the same consistency. For steamboat, house and other inside new work the following mode of operation is recommended :

First give the work a coat of shellac varnish. Make this of bleached shellac and alcohol, in proportion of two pounds of shellac to one gallon of alcohol, which must be what is known as ninety-five per cent. alcohol; because, if of less strength, it will not dissolve the gum. Then apply a coat of pure white lead, mixed in the ordinary way. Let this coat be a heavy one, and be sure that the lead is pure. Let this get thoroughly dry ; then sandpaper, and apply another coat of shellac varnish. Now, if you propose to produce a gloss finish with as little expenditure of time and material as possible, proceed in this wise: On the second coat of shellac varnish put a good, free-working, flowing coat of *pure white lead.* (If the work is to be finished in colors, this coat should be tinted to as near the finishing color as possible; and our *best* ground colors will be found cheapest for this purpose.) Supposing the job in question to be finished in white, you will mix the coat just mentioned in as little oil as possible, but enough to make it dry with a half gloss. When thoroughly dry, sandpaper well and give a coat of New White thinned with turpentine. This coat wants to flow well, and as much paint should be put upon the work as will stay there without running. This should stand a day or two to harden, and the finish or gloss coat should be applied under the same directions as given for the preceding coat. After one night the work will be fit to use.

Going back to the work as it stood with the second coat of shellac varnish, and supposing it to be the intention to finish with a perfect surface, after the style of coach panels, the mode of proceeding will be as follows : On the varnished surface put a heavy coat of our prepared rough-stuff, which is white (and which, if required, may be colored or shaded to suit the proposed finish color). This should stand one or, better, two days, before the second coat, which should stand three days before smoothing, which should be carefully performed with lump pumice stone for flat places, and good No. 2 sandpaper for beads and mouldings, and care must be taken not to rub off the corners and high places. Use plenty of water in the pumicing and keep the corners of the stone well filed. The rubbing should, of course, be followed by a thorough washing in clean cold water, the corners well cleaned out with a water brush, and when dry the surface will be ready for the coat preceding the gloss. From this to the finish the process will be the same as before described. If, after a day or two, there shall not appear to be sufficient gloss to suit, an extra coat of gloss may be put on; but to make a first-class job the gloss on the previous coat should be rubbed off with finely ground pumice stone and water. By following the above directions, a job may be done which will last almost a life-time. There will be neither tack nor crack. The material is sufficiently elastic for outside work, and hard-drying enough for work which is to be handled inside. For work in parti-colors, the most beautiful results may be produced by the adoption of the new process. It should be borne in mind that only our coach painters' colors are suited for making tints in this style of painting; common oil colors will not do.

And now our little book has reached its allotted limit, and must come to an end. With thanks to those who have pa-tiently read what we have written, hoping all may have found some instruction in its pages, asking due acknowledgement of its merits, if it have any, and leniency in judging its short-comings and defects, we bid you all, readers, a respectful Good-bye!

FINIS.

www.ingramcontent.com/pod-product-compliance
Lightning Source LLC
Chambersburg PA
CBHW021819190326
41518CB00007B/662